PESTS, DISEASES AND BENEFICIALS

Friends and Foes of Australian Gardens

F. David Hockings AM

CSIRO

PUBLISHING

National Library of Australia Cataloguing-in-Publication entry

> Hockings, F. D. (Francis David) author.
>
> Pests, diseases and beneficials : friends and foes of Australian gardens/by F. David Hockings.
>
> 9781486300211 (paperback)
> 9781486300228 (epdf)
> 9781486300235 (epub)
>
> Includes index.
>
> Garden ecology – Australia.
> Plant diseases – Environmental aspects – Australia.
> Beneficial insects – Environmental aspects – Australia.
> Garden pests – Environmental aspects – Australia.
>
> 632.0994

Published by
CSIRO Publishing
36 Gardiner Road, Clayton VIC 3168
Private Bag 10, Clayton South VIC 3169
Australia

Telephone: [+613] 9545 8555
Email: csiropublishing@csiro.au
Website: www.publishing.csiro.au

Front cover: Hibiscus caterpillar (top), bacterial spot on *Brunsfelsia* (middle), brown predatory bug (bottom)

Back cover: Bees in flowers (left), bronze orange bug (right)

Set in Palatino 11/14 and Myriad Pro
Edited by Joy Window
Cover and text design by Andrew Weatherill
Typeset by Desktop Concepts Pty Ltd, Melbourne
Index by Indexicana
Printed by Ingram Lightning Source

Feb26_RP_ILS

CONTENTS

ACKNOWLEDGEMENTS..V

INTRODUCTION...VII

CHAPTER 1   FRIEND OR FOE?   1

CHAPTER 2   CLASSIFICATION – THE IMPORTANCE OF SMALL ANIMALS   3

CHAPTER 3   PLANT DAMAGE – BY MOUTHPARTS, EGG-LAYING, BITES AND STINGS   23

CHAPTER 4   PLANT DISEASES AND USEFUL, HARMLESS AND BENEFICIAL ORGANISMS   27

CHAPTER 5   DEALING WITH PESTS AND DISEASES   31

ILLUSTRATIONS I   PESTS ASSOCIATED WITH FLOWERS, FRUIT, SEEDS, LEAVES AND SHOOTS   35

ILLUSTRATIONS II   LESS HARMFUL, HARMLESS AND BENEFICIAL SMALL ANIMALS   105

ILLUSTRATIONS III   PESTS ASSOCIATED WITH TWIGS AND SMALL BRANCHES   151

ILLUSTRATIONS IV   PESTS AND BENEFICIAL ANIMALS ASSOCIATED WITH TRUNKS AND
                       MAIN BRANCHES   165

ILLUSTRATIONS V   PESTS AFFECTING COLLARS AND ROOTS   175

ILLUSTRATIONS VI   SMALL ANIMALS ASSOCIATED WITH SOIL, COMPOST AND SHELTER   187

ILLUSTRATIONS VII   FREE-RANGING SMALL ANIMALS   205

ILLUSTRATIONS VIII   DISEASES AFFECTING GARDEN PLANTS   219

ILLUSTRATIONS IX   HARMLESS AND BENEFICIAL OTHER ORGANISMS AND SPECIALISED
                       ROOTS AND BARK   233

ILLUSTRATIONS X   PARASITIC PLANTS   243

ILLUSTRATIONS XI   PHYSIOLOGICAL DISORDERS   247

ILLUSTRATIONS XII   HORTICULTURAL PROBLEMS   257

APPENDIX I: CONTROL CHART...261

APPENDIX II: GLOSSARY...263

INDEX..265

ACKNOWLEDGEMENTS

First I acknowledge my mentor, Percy Francis Hockings – a strange relative (great grandfather's nephew), who was an architect of some note, artist and Shakespearian scholar. He became disillusioned with society, fled, first to Fiji, then Captain Cook's Possession Island off the tip of Cape York, Queensland, Celebes, finally in 1912 becoming curator of the Hockings Museum on Thursday Island, Queensland.

When I was a child, anything of interest the family members found (animal, mineral, vegetable) went to Uncle Percy and later he would send a painting and anything he could find out about it. It was a wonderful way to grow up with a mind enquiring 'What is it?' and 'What does it do?' – which is the basis of this book.

I reiterate the acknowledgements in the first edition, of now past colleagues in the Department of Primary Industries (DPI) – Brian Cantrell, Ken Houston, John Alcorn, John Donaldson and Ian Galloway – who greatly assisted with identification of many specimens. I also acknowledge Lance Warrell and Merv Whitehouse of the DPI Soil Chemistry Laboratory, who patiently enlightened me about the world of soil chemistry, plant nutrition and physiological disorders, Ian Hughes, Plant Pathologist, the master of plant problem diagnosis and

Bob Colbran, Nematologist, who introduced me to the world around the roots of plants.

I am deeply indebted to my wonderful family for their encouragement and assistance with preparation of this manuscript. I acknowledge my wife, Olive, for support and proofreading, and in particular my daughter, Cecily, and son-in-law, John Daniels, for their extensive help with the computer and the solving of photographic problems of scanning, storage and retrieval. Cecily has been a tower of strength through a major health crisis. Daughter Lindy I particularly thank for very capably helping out with drawings when photographic subjects were not immediately available; and my son, Colin, for his help in locating difficult photograph subjects.

Finally I acknowledge Bill Molyneux for his patience and persistence, which ensured that finally the manuscript was completed.

Photographic/art credits:
Bob Luttrell 189c, 191b, 192
Olive Hockings 113, 231, 342
Lindy Wieland 123c, 237, 259, 274, 328b, 328c
Colin Hockings 193b, 193c, 283a, 327, 344
John Daniels 368
Percy Hockings 65a

INTRODUCTION

I was inspired to write this book in the 1960s with the evolution and elevation of 'gardening' to amenity horticulture, the advent of television (at first only black and white) and the beginning of the development of modern pesticides, which replaced nicotine sulphate, arsenate of lead, kerosene and soft soap, and Bordeaux spray, then the main preparations in use.

On 2 January 1960 I began an appointment with the Queensland Department of Agriculture and Stock (later to become DPI). The then Director of Horticulture, Dr Sandy Trout, had the far-sighted idea that ornamental plants would become an important industry, and it was because I had a botanical bent towards plants that I was appointed for this aim. The appointment was strongly resented by the other branches of the department (due to competition for funding) and I was occasionally approached by other branch officers and told off in severe terms.

Garden clubs, horticultural societies and more specific organisations such as orchid societies were flourishing and speakers were in high demand. It came about that I shared a stage with a highly respected garden commentator who, when presented with some much chewed azalea leaves, announced with authority that this was the result of severe lace bug attack. As bugs have piercing and sucking mouthparts, they cannot take even the tiniest bite out of a leaf. I was hard pressed to sit calmly.

This general lack of basic understanding was quite obvious in the television programs of the time, with presenters continually recommending the 'magic' new pesticide Rogor™ for fungal, pest and all other plant disorders. Quite obviously the professionals of the day, as well as the gardening public, were struggling to understand, identify and classify the many small garden creatures. Containers of pesticides and their advertising did not differentiate between pests, harmless or beneficial creatures – the attitude was 'if it moves, spray it'. I committed myself to doing something about this.

In this book I have attempted to provide illustrations of a comprehensive cross-section of the various small creatures that can be found in the average garden, to serve as a general guide to all garden and plant situations everywhere. Equivalent species with similar life cycles and feeding habits occur throughout this country and with, in some cases, climatic modifications, also in overseas countries.

In order to examine and identify small animals, a magnifying glass (×10 or ×20) is essential. The small mites, such as the broad mite, cyclamen mite and brevipalpid mite, require ×20.

For easy reference I have attempted to group creatures according to where on plants or in the surroundings they are most likely to be encountered: leaves and shoots, stems, trunks, soil, compost, and so on.

Unfortunate and confusing colloquialisms such as referring to all small creatures as 'bugs' or using the term 'worms' for worms, caterpillars and maggots, has largely been avoided. These terms are more correctly used for the specific groups to which the name applies – 'bug' for the Hemiptera (the real bugs, which have piercing and sucking mouth parts), 'worm' for Aschelminthes and Annelida (earthworms and nematodes) and 'maggot' for the larvae of wasps, bees, ants (Hymenoptera) and flies (Diptera).

I have applied my own idiosyncrasy of reserving the name 'caterpillar' for the larvae of moths and butterflies (Lepidoptera) and 'grub' for the larvae of beetles (Coleoptera).

Since this book was first published in 1980, there have been major changes in the registration and range of pesticides used for control of pests and diseases and the protection of cultivated plants. I have attempted to cover these changes in Chapter 5 and Appendix I.

This republication has afforded me the opportunity to expand the subject material, better control of photographic reproduction, better clarification and cross-referencing and, I trust, better connectivity in an easier-to-use book. It is with some satisfaction that it is now completed.

David Hockings
August 2013

Chapter 1

FRIEND OR FOE?

One of the really absorbing interests associated with gardening is the discovery of the large number of weird and wonderful insects and other small animals that have evolved with plants. They are not all pests and some are quite beautiful or else have a bizarre appearance.

Apart from the native creatures, a large number of introduced species, which include many pests, have become established here in Australia.

Some of these small native and exotic animals (they are not all insects) are quite specific to particular families of plants, some even to particular species, while others have a wide range of association. The larger creatures are usually fairly obvious, although some are remarkably well camouflaged.

Small insects such as aphids and planthoppers are not easily noticed, and the damage or abnormality they cause is more likely to attract attention before the creatures themselves are detected.

Still other creatures, such as mites, are so small that, even though you may look closely, you are unlikely to see them.

Smaller mites such as broad mite and brevipalpid mite often do damage to the growing point or young fruit or the bark of young stems – damage which does not become evident until, after a period of growth, the damaged tissue tries to expand. By the time the damage is noticed, the mites may no longer be present. I call these cyclic mites.

Some pest damage is easily confused with damage caused by disease organisms (fungi, bacteria and viruses) or abnormalities from physiological causes. Physiological abnormalities are many and varied and may result from such factors as deficient or excess trace elements, sunburn, over-fertilising or accumulated salts. Herbicide damage has become very common and can also be confused with pest damage.

Garden pest and disease control chemicals are almost all quite specific for either pests or diseases (unless 'catch all' mixtures). Products specifically for use against small animal pests are, in almost all cases, quite useless against diseases and vice versa. Some are fairly ineffective against other than a particular group of pests or particular group of disease organisms. For example, sulphur has some effectiveness against mites as well as rust fungus, but is of little use against most insects or other fungal diseases or bacteria or viruses. Miticides are used against mites, but are of little use against insects.

It is pointless spraying or dusting with pesticide unless you know what pest you are dealing with, and it is essential to know if treatment is even necessary. It should be understood that not all of the small animals that can be found in gardens are pests. Many of them are harmless or beneficial in a variety of ways.

The beneficial and harmless creatures usually outnumber those that actually damage plants. Even of those that feed on plants, a significant number do so little damage that treatment or removal is rarely warranted, and they may have other virtues.

If a pesticide spray or dust is avoidable, the environment and your plants would be better off without it. I would never want a garden without insects – just imagine, no butterflies or bees!

Insects, like the nectar-feeding birds, have evolved along with the flowering plants. Bees, wasps, flies, beetles, moths and butterflies play a significant part in pollination and seed production of crop plants as well as bush plants.

Insects and the other small animals also play an extremely important part in the disintegration and decomposition of plant debris, animal wastes and the bodies of dead animals. This action is extremely important in disposal of litter, recycling of plant nutrients, and formation and improvement of soil.

It is therefore obvious that some understanding of the activities and feeding habits of the various small animals is most important if you are to identify whether they are harmful, harmless or beneficial.

A **predator** is a carnivorous organism that either captures and consumes its active prey or, more passively, consumes soft-bodied, relatively inactive prey such as aphids. A predator may have raptorial forelegs for seizing prey and sharp mouthparts for tearing the victim apart, or it may have a piercing style through which it injects a digestive fluid that predigests the contents of the victim – which are then sucked out.

Some predators use snares (spiders) or traps (ant lions), and some wasp predators capture and anaesthetise prey to provide food for their larvae.

A **parasite** is an organism that takes or steals its nutrition from another organism (the host). Its parasitism usually leads to death of the host.

A parasite can be classified as an **ectoparasite** (it feeds from outside the host), **endoparasite** (it feeds within the host), **hyperparasite** (a parasite on a parasite) or **parasitoid** (it does not kill its host – eggs are laid amongst or in the host eggs, or on food that may be eaten by the host).

As an aid to this understanding, the following two chapters show how these various small animals are grouped, the way the groups are related to each other, and how the members of these groups feed and multiply. It is important to realise that even within genera of pests there are frequently species that are parasites or predators of the pest species.

Chapter 2

CLASSIFICATION – THE IMPORTANCE OF SMALL ANIMALS

In times past, the world's living entities have been classified into either the Plant Kingdom or the Animal Kingdom. However, some forms of bacteria, protozoa, fungi, slime moulds, water moulds and algae have exhibited some characteristics of both these kingdoms.

To deal with these anomalies five or six kingdoms are now recognised: Plantae, Animalia, Fungi, Protista and Monera. Monera is further divided into Eubacteria and Archeobacteria. Some of these changes relate to plant disease-causing organisms and others that can be found in gardens.

Kingdom Protista now includes the slime moulds (myxomycetes), water moulds (phytophthoras), and human disease organisms such as *Amoeba, Plasmodium* (malaria) and *Giardia*.

Kingdom Fungi contains food species such as mushrooms, truffles, morels, and useful species such as yeasts (bakers and brewers), penicillins (cheese, yoghurt and medicinal), and mycorrhizal and entomogenous fungi. Also there are many plant disease species as well as human-disease organisms such as *Cryptococcus* and *Candida*.

Kingdom Monera includes the bacteria and is sometimes divided into kingdom Eubacteria and kingdom Archaeobacteria.

Kingdom Eubacteria includes important decomposers and recyclers or symbionts in the gut of many animals to aid digestion (in termites, even in man). A couple that cause human diseases are *Escherichia coli* and *Salmonella*.

Kingdom Archaeobacteria includes species that live in extreme environments such as salt lakes, hot springs and undersea volcanic vents.

In the Animal Kingdom, the levels of classification are as follows: phylum, class, order (sometimes further divided into suborder and superfamily), family (sometimes subfamily), genus and species (sometimes further into subspecies and variety). At times taxonomists have rather flexibly used another level 'higher taxon' as an intermediate between main levels of classification.

The classification into groups called phyla (singular, phylum) applies to the simplest forms of organisms, the microscopic single-celled animals at the bottom of the evolutionary ladder, and equally to organisms of increasing complexity and evolutionary development, including humans.

The divisions of classification and the phyla into which some animals are placed vary slightly according to which authority is followed. According to some systems of classification, there are some 22 phyla of animals, but fortunately we are concerned here with only four phyla and with relatively few of the individual small animals in these phyla.

PHYLUM NEMATODA

NEMATODES OR ROUNDWORMS
[ILLUSTRATIONS 6, 108, 247]

This group was previously included in the phylum Aschelminthes. It includes the nematodes and consists of slender unsegmented cylindrical or sac worms, mostly small to microscopic in size. They reproduce by laying eggs and the sexes are usually separate. The young worms moult (grow a new cuticle or skin beneath the old, which is shed) as they grow.

Aschelminths include many marine and freshwater worms and many terrestrial and parasitic species such as the roundworms and hookworms, as well as nematodes and the *Gordius* or horsehair worms.

Nematodes are thread-like worms, although the mature females of some species are swollen or sac-like. They include many that are parasitic on or in other forms of animal life, including humans, and many that are parasitic on plants.

The majority of soil nematodes are microscopic, ranging in length from 0.5 mm to 2 mm. Different groups of nematodes feed in the following ways: as parasites of insects and other small animals (often of the larval stages); as predators on other nematodes; as feeders on fungi, bacteria, algae and moss; and as parasites of plants,

causing loss of vigour, decline disease and, in some cases, spread of virus diseases. Many of the serious plant pest species are introduced to Australia.

Parasitic nematodes that attack plants *below* the ground may cause root galls, root rots, root lesions, damage to the surface or tips of roots, distortion or bunching of roots.

Parasitic nematodes that attack plants *above* the ground may cause distortion or death of growing points or flower buds, leaf or flower bud drop, galls or angular leaf spots.

The natural enemies of nematodes include: parasitic protozoans (single-celled animals) as either external or internal parasites; predacious nematodes that may swallow pest nematodes whole or puncture and suck out the contents; and parasitic and predatory fungi that either parasitise or actively capture nematodes. Other groups of worms as well as some mites and springtails also feed on nematodes.

Nematodes are of very great economic importance, perhaps even greater than is generally realised. There are many groups besides the more obvious root-galling (root knot) types. Being microscopic and mostly hidden under the soil, they are easily overlooked and the symptoms are often attributed to other causes such as root-rotting fungi or faulty nutrition. Some historical rural catastrophes such as the American 'Dust Bowl', recorded as having other causes, are now believed to have been the result of nematode infestation.

Nematode problems are best avoided by strict rejection of infested plants and soil. Expensive and very poisonous chemicals such as fenamiphos (Nemacur™) can be used to treat infested soil but kill other beneficial soil animals, can be harmful to plants and, if used close to harvest of edible plants, can render the plant parts poisonous. Leaf nematodes may be controlled by pruning and shifting plants to a drier environment.

Attempts at biological control of nematodes, using predatory species or nematode-feeding fungi and trap plants such as marigolds (which attract nematodes but in which they cannot reproduce), have been disappointing.

The best long-term solution is regular addition of organic matter as mulch. Organic matter promotes a large array of organisms that assist in decomposition and recycling nutrients. Some of the fungi and bacteria feed specifically on nematodes. A DOOR ('Do Our Own Research') trial carried out with hardwood sawdust mulch took about 10 months to be colonised by nematode-feeding fungi and bacteria and then gave complete control of nematodes.

The sawdust was pre-treated to prevent nitrogen drawdown by mixing in urea and moistening, then composting for six weeks. The mulch layer had to be topped up every six months to maintain the layer.

Phylum Mollusca

This phylum of **slugs** and **snails** consists of soft-bodied animals, some of which have a single, bivalve or compound shell. They are oviparous (reproduce by eggs) and the sexes are usually, but not always, separate.

There are several classes of molluscs, most of which are marine animals such as chitons, oysters, clams, tooth shells, conch shells, cuttlefishes, octopuses, squids and nautiluses.

We are concerned here with only slugs and snails of which there are both marine and terrestrial species.

Class Gastropoda (snails, slugs, conch shells)

These animals may have a single shell (usually coiled) or no shell. Of relevance to this book are the terrestrial snails and slugs whose feeding habits include the following: predators on other animals including worms and other snails and slugs; feeders on decomposing vegetable material, compost and leaf litter; feeders on fungi, algae and mosses; and feeders on higher plants including cultivated plants, by rasping and swallowing tissue.

Snails [Illustrations 201, 202, 248, 249]

A large number of native Australian land snails occur throughout the country, with the largest concentration of species and the largest species being found in the tropical and subtropical rainforests of Queensland and northern New South Wales.

There are also some 15 or more introduced snails established in parts of Australia and gardeners are in fact more familiar with the introduced species (some of which are destructive garden and farm pests) than with the native snails. The native snails can rarely be found living outside their natural bush environment.

There are several predatory species, both introduced and native, which eat other snails, slugs and other small animals.

Slugs [Illustrations 250, 272]

Australia apparently does not have a large number of native species of land slugs, and in any case the native slugs are rarely seen in gardens. As with snails, about 10 or more introduced slugs (including some that are serious pests of gardens and farms) are the ones usually to be found in gardens. The native slugs, some of large size, mostly feed on plant and animal refuse and forest litter, algae or lichens. One of the introduced slugs is carnivorous, feeding mainly on worms.

Pest species of snails and slugs can be controlled by baiting with pelleted preparations of methiocarb or metaldehyde, which are available commercially.

Phylum Annelida

This phylum contains the segmented worms such as **earthworms**, freshwater and terrestrial leeches and some of the marine species such as bristle worms and sand worms.

The annelids reproduce by eggs and some also asexually by division. The sexes may be separate or the animals may be hermaphrodite but with eggs that require fertilisation by another individual. The young emerge as small but fully formed worms.

Class Oligochaeta (earthworms) [Illustration 271]

There are several native earthworms but not as many as can be found in non-arid countries. Here, other creatures such as termites tend to be the main recyclers – they are not all destroyers of timber. Some of the native earthworms are of particular interest because of their very large size, one recorded to 3.5 m in length and another to 1.2 m in length. They are very rarely seen because in gardens and cultivated areas they have been replaced by accidentally introduced European and Asian species – ferals.

Earthworms feed mainly on fallen leaves and other vegetable litter, which they take down into the soil. A worm's digestive system includes a muscular gizzard in which vegetable matter and soil are finely ground. Worms make extensive tunnels through the soil and absorb nutrients, which are recycled by being deposited on the surface outside the burrow as castings.

Earthworms have gained the reputation of being the most beneficial animals in the world because of their activity of aerating and mixing soil. However, they can be troublesome pests of potted plants (by breaking down the potting mix and clogging drainage holes) and of lawns and greens (because of their castings). On a few occasions they have been found to be so numerous and active that they prevented plants from growing.

Earthworms are preyed upon by slugs, springtails, beetles and birds and parasitised by protozoa, bacteria, mites and flies.

PHYLUM ARTHROPODA

This phylum contains the great majority of the small garden animals such as **slaters**, **centipedes**, **millipedes**, **mites**, **spiders**, **earwigs** and all of the **insects**. The adults of the arthropods have hard outer skeletons, segmented bodies and paired, jointed appendages. They are usually oviparous (reproduce by eggs), rarely viviparous (bearing live young) and the sexes are usually separate.

There are many classes of arthropods comprising an immense number and variety of animals, some minute. Arthropods of some kind can be found in every possible situation (in the sea and freshwater, on land, in the air and in the soil). Many feed on plants and many feed on other small and large animals as either predators or parasites. Many are pests of livestock and man. Some classes of arthropods contain species that are of great importance directly or indirectly (as vectors of disease) to crops and human health.

SUBPHYLUM CRUSTACEA (SLATERS, AMPHIPODS, LOBSTERS, SHRIMPS, CRABS, CRAYFISH, PRAWNS) [ILLUSTRATIONS 264, 265]

Most of these are marine animals and not relevant here, but some live in freshwater or damp places on land. The slaters or wood lice and the flea-like amphipods are the only members of this class to be dealt with here.

Slaters are very secretive little animals, hiding under stones, timber, pots and so on, and feeding mostly on decaying vegetable matter. However, occasionally they are minor pests, feeding on young seedlings and cultivated mushrooms.

Amphipods look like large fleas (or minute prawns), and in fact jump like fleas. They are common and important in compost heaps and decaying vegetable matter, breaking it down to humus. There are instances of numbers of amphipods entering low-set homes in wet weather and subsequently dying and causing an unpleasant smell.

SUBPHYLUM MYRIAPODA

Centipedes [Illustrations 267, 268]

A few centipedes are littoral species, but most are land-dwellers living under stones, timber, bark, etc.

They are capable of biting and have venom glands, but many species are probably more repulsive than dangerous. They are slightly beneficial because they feed on other small animals, some of which are pests of cultivated plants.

Millipedes [Illustrations 269, 270]

All are land-dwelling animals which, like centipedes, live under cover.

They have stink glands instead of venom glands and are quite harmless and relatively inoffensive. Generally they are slightly beneficial in that they help reduce plant litter to humus. Some species have been known to damage potatoes and other crops but usually they are of little importance in Australia.

SUBPHYLUM CHELICERATA

Class Arachnida (scorpions, false scorpions, mites, ticks, spiders)

This class of small animals is almost as varied as the crustaceae, but they are almost all land animals. Their importance to plants and gardening is very much greater than that of the crustaceae, but they are not as important as are the insects.

The arachnids are often confused with insects but differ from insects in normally having eight legs (instead of six), a body of two parts (instead of three), no wings and no antennae. Their life cycle consists of egg, several sizes of miniature immature young and then adult.

Arachnids are divided into about 10 orders (depending on which authority one follows) and about five of these are relevant to the average garden.

Order Scorpiones (scorpions) [Illustration 231]

Scorpions can sting with their 'tails', some quite painfully, and they are regarded with some dread. Most of the Australian species are probably more repulsive than dangerous, but should be treated with caution.

They live a rather secretive life, hiding under stones and timber and in other crevices or in soil burrows, but are more active at night. The mature female gives birth to tiny immature young which, for a time, ride on their mother's back. Scorpions are beneficial animals in that they feed on other small animals, including plant pests.

Order Pseudoscorpiones (pseudoscorpions) [Illustration 232]

These small scorpion-like animals look very like true scorpions, but lack the tail and sting and are quite harmless.

They are sometimes encountered under loose bark or pots. They are probably slightly beneficial in that they may feed on pest animals.

Higher taxon acarina (mites and ticks) [Illustrations 11, 53 to 59, 195, 196, 212, 235, 263]

There are many different families of mites, including many that are very serious pests of plants. Others are scavengers and detritus feeders, feeding on decaying vegetable and animal debris. Mites feed by piercing and sucking. Some are pests of stored food (cheese, dried meat, flour, seeds, dried fruit and plant bulbs). Fortunately there are also beneficial mites, species that prey upon or parasitise other mites, insects and other small animals.

Order Trombidiformes (detritus-feeding, plant-feeding and predatory mites)

Plant-feeding mites can occur in enormous numbers and seriously damage plants. The main families of the most troublesome mites are:

- superfamily Tetranychoidea – **spider mites** (red spider mites, two-spotted mites, and so on), the adults of which are just large enough to be visible.

- family Tenuipalpidae – **brevipalpid mites, passion-vine mites, false spider mites**. These are very tiny and require ×20 magnification.

- superfamily Tarsonemoidea – **broad mites, cyclamen mites**. These colourless mites are very small and require ×20 magnification to be seen.

- superfamily Eriophyoidea – **erinose mites (blister or gall-forming mites)**. Most are even smaller than the above and invisible without a microscope. They are mostly torpedo- or thread-shaped. The often misidentified tomato mite is one of these.

- superfamily Trombiculoidea, family Trombiculidae – **(harvest mites)** [Illustration 263]. These small animals are generally larger than other mites and can sometimes be found in compost heaps and leaf litter. Some are harmless creatures that feed on decaying vegetable material, and some species are predators of other tiny animals.

Order Mesostigmata

Superfamily Phytoseioidea, family Phytoseiidae (native **predatory mites**). These are usually fast moving mites that prey on other mites.

Adult female mites lay tiny spherical eggs on the surface or in crevices of the host plant. These hatch into minute six-legged mites, but the later stages have eight legs. Dry weather often favours mite infestation and likewise the plants growing under the shelter of the eaves are frequently the worst infested.

The natural enemies of mites include predatory mites, ladybird beetles, lacewings, hoverflies, tiny wasps, lizards and fungi.

Mites are not insects, so frequent application of insecticides tends to increase trouble with mites simply because insecticides kill off the insects that feed on mites. Wettable sulphur or even dusting sulphur may be sufficient to control light infestations, but severe infestations will require application of a miticide.

Order Ixodida

Superfamily Ixodoidea, family Ixodidae – there are **ticks** (including two introduced species) that are pests of livestock, and other ticks (or are these large mites?) can sometimes be seen on reptiles and insects. The tick of particular concern is *Ixodes holocyclus*, the paralysis tick. It is occasionally a problem in gardens that are frequented by bandicoots or possums (particularly if adjacent to unmown grass or bushland). These native animals play a part in the tick's life cycle. When fully engorged, the female tick falls off the host animal and becomes virtually a sac of eggs. On hatching, the minute ticks climb to the top of leaves of grass or shrubs where they wait to be brushed on to a passing animal. They grow through several moults. They should be pulled off immediately when they are detected – without any treatment (which will make them inject toxin). A roll-on deodorant or insect repellent can be applied to the groin, armpits and around the waist to repel attacks.

Order Araneaea (spiders) [Illustrations 167 to 172, 230, 259 to 262]

Spiders are predators of insects and other small animals, sometimes including other spiders, and very rarely of frogs, small birds, lizards and even mice. They are an extremely beneficial group, snaring, trapping, ambushing and hunting, mainly at night. Their special value is that they capture many egg-laden adult pests.

Most people regard spiders with a great deal of dread. Spiders are generally listed or illustrated on cans of, or advertisements for, pest destroyers, along with cockroaches, flies, fleas and so on. Over 1500 species of spider have been recognised in Australia so far, and of these only about 18 native species and some four introduced species can in any way harm humans.

Spider webs on buildings can be an unsightly nuisance, and also the occasional tight webbing of twigs on fruit trees and garden plants can prevent normal growth. However, in general they are seriously misunderstood, highly beneficial small animals that make extensive use of silk (or web) in quite remarkable ways.

Spiders and their eggs are preyed upon by birds, lizards, frogs and predatory insects, and parasitised by wasps, flies and some moths.

Subphylum Hexapoda

Class Entognatha

Order Collembola (springtails) [Illustration 258]

These are minute, more or less colourless or brown, wingless insects that can spring a considerable distance in relation to their size. Some species of springtail are very common inhabitants of compost heaps and garden soil and can be found under pots and stones. Their food consists mainly of decaying vegetable and animal matter, fungi, algae and lichens.

In general they are slightly beneficial in that they help the breakdown of waste to humus, but they can be pests of mushroom cultivation and sometimes the roots of crop plants. One is a serious pest of lucerne and other leguminous crops.

Springtails are eaten by mites, beetles and other small animals.

Class Insecta (= Ectognatha) (insects)

This class of animals is by far the largest and most varied of all. It contains the majority of the small animals that may from time to time be seen on or near plants.

The insects differ from the other arthropods in that the adults have six legs,

a body of three parts, antennae and, in the adults, usually wings. There are two types of life cycle depending on the order to which the insect belongs:

1 (complete metamorphosis) egg, larva (caterpillar, grub or maggot), pupa (chrysalis), adult (examples are moths, butterflies, beetles, flies, wasps, bees and ants).

2 (incomplete metamorphosis) egg, several nymph stages (wingless miniatures of the adult), adult (examples are bugs, aphid leafhoppers, stick insects, mantids, cockroaches and termites).

Insects are divided into many different orders and, of these, 17 orders contain species that are frequently to be found in average gardens.

Order Odonata (dragonflies, damselflies) [Illustrations 284, 285]

This is a group of predatory insects and, as such, they are beneficial. The nymphs are almost all aquatic and they feed on other small aquatic animals. The adults are very active predators, capturing other small insects in flight.

Although they breed in water, the young adults can be seen in gardens a considerable distance from water. However, after the young adults have developed full mature colouring, they move back to water for breeding.

The aquatic nymphs are eaten by fish, frogs, birds and reptiles, and the adults by robber flies, mantids, bugs and wasps as well as spiders. They are also parasitised by worms.

Order Blattodea (cockroaches) [Illustrations 200, 254]

This very distinct group of insects is best known by the nine introduced pest species which destroy and pollute food. Of the many (over 400) native species, few ever enter houses and scarcely qualify as pests.

The native species are quite harmless and feed on fallen leaves, rotten wood or forest litter. The very large, blind 'Moonie' cockroach grows to about 5 cm in length, lives in soil burrows and feeds exclusively on fallen eucalypt leaves. It makes an interesting pet. One native species often seen on plants could be beneficial because it includes the adults of fruit flies and flea beetles in its diet – but perhaps it eats only dead or dying insects.

Cockroach eggs are laid in a cluster enclosed in an ootheca or case. The nymphs are miniatures of the adult but often differently coloured.

The eggs are heavily parasitised by small wasps and beetles. The nymphs and adults are eaten by birds, frogs, reptiles and marsupials (bandicoots, marsupial mice) and predatory arthropods. They are also parasitised by several species of worms, protozoa and amoeba.

Order Blattodea, previously Isoptera (termites or white ants) [Illustration 246]

These small soft-bodied insects live secret lives hidden away in galleries or tunnels beneath the ground or in clay mounds constructed above the soil surface. Others live in galleries in tree trunks or in nests attached to trees. They differ from true ants (Hymenoptera) in that a termite, on hatching from the egg, is a miniature of the adult and grows by a series of moults. By contrast an ant, after hatching, goes through maggot and pupa stages before becoming an adult. Termites are more closely related to cockroaches than to ants.

A termite colony is highly organised into several different castes of different appearances and functions. Usually only two types of individuals (the queen and males) are permitted to develop sexually and the rest are modified into wingless, blind, sterile workers or soldier castes. Mature colonies periodically produce winged male and female forms (alates),

which are released in a swarm when weather conditions are favourable, to mate and form new colonies.

The different species of termite feed on forest litter, grass, fungi and the centres of live trees (sometimes fruit and ornamental trees). Some are serious pests, attacking construction timber, furniture, textiles, plastic, paper and many other materials used by man. On the other hand, particularly in arid countries, termites outrank earthworms in soil formation, disposal of forest litter and recycling of nutrients.

The winged adults are eaten by lizards, snakes, frogs, birds, bats, marsupials, ants, dragonflies and other predatory insects and spiders. Colonies are attacked by echidnas, marsupial mice, bandicoots, snakes, geckos and frogs.

Residual chemicals and physical barriers provide some protection, but there is an ongoing need for inspections to ensure that termites have not invaded houses. If you are unlucky enough to find a termite infestation in your house, do not interfere with the nest and call in professional controllers immediately.

Order Mantodea (praying mantids) [Illustrations 145, 154 to 156, 166]

These very distinct insects are all predators that feed on other insects and spiders, which they capture and hold in their raptorial forelegs and tear apart with their mouths.

In spite of widely held theories about the value of mantids in controlling garden pests, their economic value is very doubtful. They are not selective feeders, but will seize any small creature that happens along, including other mantids. This behaviour usually results in few mantids in a garden. The female often devours the male during courtship.

The adult female lays the eggs in a compact cluster enclosed in a frothy or spongy material that is moulded into a shape characteristic of the species. Some species attach the eggs to rocks, logs or tree trunks, others to twigs, grasses or herbs.

Tiny ant-like nymphs emerge and as they develop and grow they moult several times, finally becoming adults. Males are always fully winged, but the females of some species may be either short-winged or wingless. Many species live on trees, shrubs, grasses and herbs, some on the bark of trees and some on the ground.

The natural enemies of mantids include birds, lizards, wasps, worms and marsupials. In addition they are parasitised by mites and worms. Mantis eggs are usually heavily parasitised by tiny wasps, flies and beetles, and are also eaten by crickets.

Order Dermaptera (earwigs) [Illustration 229]

This small group of elongate insects can usually be distinguished from other insects by the fact the body ends in a shape like a pair of forceps. They are nocturnal and favour sheltered, often damp places or in leaf mould or compost heaps.

Eggs are laid by the adult female which generally guards them until after hatching. The nymphs look like small versions of the adult.

Earwigs generally feed by chewing on living or dead plant and animal wastes. Some, including an introduced species, are minor pests, occasionally feeding on flowers and fruit. Otherwise they are beneficial because they include some pest insects in their diet.

Their natural enemies include birds and bats and they are also parasitised by worms and flies.

Order Orthoptera (grasshoppers and crickets)

The insects in this order are best known for their powers of jumping and if caught by hand will kick out with their spiny hind legs. They have biting and chewing mouthparts, and one group, locusts, is amongst the most destructive plant pests in the world. A few species are predators of other insects. Many species are able to 'sing' or produce sound by rubbing parts of their wings together.

Grasshoppers [Illustrations 102 to 104, 198, 199] Grasshoppers are divided into long-horned grasshoppers in which the antennae are usually longer than the body, short-horned grasshoppers (which include the locusts) in which the antennae are less than half of body length, and a few unusual small species sometimes called grouse locusts.

Nearly all the species have the rear legs adapted for jumping and a few have the front legs adapted for burrowing or, in the case of predatory species, for seizing prey. Adults of most species are fully winged in both sexes, but there are a few short-winged or wingless species or forms. Often wing modification is confined to the females.

Some species live and feed on shrubs and trees, and others live on the ground, feeding on grasses, herbs, forest litter or algae-covered mud (grouse locusts). A small number are semi-aquatic.

Eggs are usually laid in the soil, but some species cement the eggs in rows attached to twigs. The several nymph stages look like small wingless versions of the adult. In general, the adults live a solitary life, but some species such as locusts become gregarious and swarm in enormous numbers.

Natural enemies of grasshoppers include birds (particularly ibises), mammals (such as *Antechinus* and sugar gliders), reptiles and other insects such as large wasps and assassin bugs. In addition, they are parasitised by small wasps, flies, moths, mites, worms and fungi. Wasps also parasitise the eggs.

Many species (particularly the locusts), damage crops and garden plants. There are a few species that prey on other insects.

In the garden situation, periodic collection and destruction by hand is often sufficient, but insecticides are available.

Crickets [Illustrations 197, 255, 256] Most crickets have the same jumping and singing powers as grasshoppers. There are both plant-feeding and predatory crickets as well as omnivorous species that scavenge on a mixed diet of plant and animal origin, such as mantis eggs and snake skins.

Many species live in the soil or under shelter. Some are cave-dwellers or live an entirely subterranean life, and consequently the legs of several species are modified for digging. A few species shelter among leaves or under bark.

The adults of many species are fully winged but there are short-winged and wingless species. Eggs are usually laid in the soil or, in the case of tree crickets, they may be inserted in twigs or leaves. The nymph stages closely resemble the adults, but are wingless.

A few species are pests of pasture and mole crickets are pests of bowling greens because of their raised tunnels along the soil surface. Pesticide control is rarely warranted.

Order Phasmida (phasmids or stick insects) [Illustrations 105 to 107, 157 to 160]

Most of the stick insects are relatively rare and are of particular interest because of their unusual appearance. All are leaf-eaters and many will feed on a variety of plants.

With few exceptions, they occur in only sparse numbers and although leaves are eaten, the amount of damage is unimportant, except perhaps if they happen to be feeding on a small plant. Their removal or destruction is rarely warranted. The exceptions are the three plague species that may occur in large numbers and defoliate trees or even forests. Phasmids are often confused with praying mantids, which are all predatory.

Phasmid eggs, which resemble hard seeds, are dropped on the ground where they eventually hatch over two or more seasons. The nymphs at first resemble ants, but they moult or shed their skins several times as they grow and older nymphs resemble the adult insect. In some species, both male and female are fully winged, and in others only the male is capable of flight – the female is either short-winged or wingless.

Only the plague species (*Podocanthus wilkinsoni, Didymuria violescens* and to a lesser extent *Ctenomorphodes tessulatus*) are occasionally damaging when they occur in very large numbers and defoliate sections of eucalypt forest.

The natural enemies of phasmids include birds, marsupials, lizards, spiders and predatory insects such as praying mantids. They are parasitised by tachinid flies and mites. The eggs are eaten by mice and parasitised by minute wasps. Because the insects occur in only small numbers, control is usually unnecessary but if it becomes necessary, gardeners can remove and destroy them by hand.

Order Hemiptera (planthoppers, spittle bugs, cicadas, psyllids, lerps, white flies, aphids, scale insects, mealy bugs)

Almost all of this very large and varied group of insects feed on plants and have piercing and sucking mouthparts. In some of the scale insects, mealy bugs and aphids, the males do not feed and in fact do not have mouthparts.

This suborder includes many insects that are serious pests. Where a large population of the pests is present, loss of sap causes damage. The toxic saliva of some species also severely damages and distorts new growth. Some species are vectors of virus diseases.

Many species excrete a sugary substance called honeydew onto leaves and other parts of plants. This causes the growth of black fungi called sooty mould. The honeydew also attracts ants and many species of Hemiptera are ant-attended. In addition, some of the species of Hemiptera cause galls on plants.

Order Hemiptera, suborder Auchenorrhyncha

Superfamily Fulgoroidea and superfamily Membracoidea, family Cicadellidae (leafhoppers or planthoppers)
[Illustrations 23, 24, 203 to 205] This is a very diverse group of small insects which hop away if touched or disturbed. Some live on leaves and others on stems of shrubs or on the bark of trees.

The eggs are laid by the adults on the host plant and all the nymphal stages as well as the adults feed together sucking sap. Some species excrete drops of water (the rain trees), others honeydew and some secrete wax filaments.

Some leafhoppers and planthoppers are of little importance, but others have toxic saliva or may spread virus diseases.

The natural enemies of leafhoppers and planthoppers include birds, marsupials, lizards, spiders, hoverflies, ladybird beetles, predatory bugs and lacewings. They are also parasitised by several tiny wasps and mites and are subject to fungal diseases.

If control is warranted, systemic insecticides are available.

Superfamily Cercopoidea (spittle bugs and froghoppers) [Illustrations 177, 178] A characteristic of this group is the need of the small, soft-bodied nymphs to keep themselves wet. The nymphs of spittle bugs achieve this by producing a blob of frothy bubbles under which they shelter and feed. Froghopper nymphs construct hard, horn-like shelters attached to twigs. These shelters are enlarged as the insect develops and nymphs remain immersed in liquid in the structures.

The adults of both are typical planthoppers, and both nymphs and adults feed by piercing the plant and sucking sap.

Their life cycle, predators and parasites are the same as for planthoppers but, as they are relatively harmless to plants, spittle bugs can if necessary be hosed off and froghoppers pruned or rubbed off.

Superfamily Cicadoidea (cicadas)
[Illustrations 193, 194, 215, 257, 329] Cicadas are small to larger insects best known for the shrill, ear-piercing noise made by adult males on warm summer days, or some species at sunset. Sometimes the name 'locust' is wrongly applied to cicadas.

The adult female is equipped with a saw-like ovipositor with which she cuts zigzag slits in the bark of twigs and small branches. Sometimes a cut so weakens a stem that it may later break off in a strong wind. Eggs are deposited in the slits and the tiny nymphs, on hatching, jump or fall to the ground where they burrow deep in the soil and suck from roots. The nymph stage may last for several years. The final nymph stage burrows to the surface and, at a suitable time and favourable weather, emerges, climbs a nearby support where it attaches itself and the winged adult emerges.

Although the nymphs suck from roots and the adults from branches, they are not generally recognised as garden pests. However, two species are important pests of sugarcane.

The natural enemies of cicadas include birds, marsupials, lizards, spiders and predatory insects such as wasps, mantids and bugs. The nymphs may be parasitised by mites and are subject to fungal disease. (See Illustration 329.)

Order Hemiptera, suborder Sternorryncha

Superfamily Psylloidea, family Psyllidae (psyllids) [Illustrations 7, 12 to 21] Psyllids are small to minute sap-sucking insects and are probably the largest and most varied group of pests of native plants. The family also includes the **lerp**-constructing species

[Illustrations 19, 20, 21] the nymphs of which construct distinct coverings or testa (lerps) on leaves or twigs in which they shelter and feed. Nearly all lerps are associated with eucalypts and some of these coverings are very elaborate and beautiful. There are also many gall-forming or leaf- and shoot-distorting species of psyllids.

Some species secrete fluffy, white, waxy material on young shoots or on the undersides of leaves. Others secrete long white waxy filaments and two species are known to live under a layer of sticky solidified sap on *Ficus* leaves. Some are free-living, while others live in colonies and may be ant-attended. Honeydew excreted by psyllids attracts ants and also causes growth of sooty mould. Some species of psyllid have toxic saliva, which causes the death of leaves but rarely the death of host trees.

The adult females lay eggs on the host plant and the nymphal stages, as well as the adults, feed by piercing and sucking. The nymphs may look like small, wingless versions of adults or, in some species, may be scale-like or otherwise modified as in the case of gall-forming species. Some cause serious disfigurement of ornamental plants.

The natural enemies of psyllids include birds, marsupials, spiders and predatory insects, particularly ladybird beetles, hoverflies and lacewings. In addition they are parasitised by minute wasps and fungal diseases.

Systemic insecticides are available to control psyllids.

Superfamily Aleyrodoidea, family Aleyrodidae (white flies) [Illustrations 25 to 28] White flies are tiny sap-sucking insects allied to psyllids. They are given the common name of white fly because of the white appearance of the tiny winged adults, which cover their bodies and wings with a fine white powdery wax secretion.

The adult female attaches eggs to leaves, some arranged in a distinctive pattern. The first-stage nymphs move about on the host plant for a short time, then attach to the leaf and are sometimes mistaken for scale insects. These later nymphal stages remain attached and immobile and finally the tiny four-winged adults emerge.

Besides several native species, there are several introduced species which are serious pests of crops and ornamental plants.

Their predators, parasites and means of chemical control are similar to psyllids.

Superfamily Aphidoidea, family Aphididae (aphids) [Illustrations 22, 240] In Australia, aphids consist of about eight endemic species specific to native plants and about 100 introduced species, most of which are pests of cultivated plants.

In most species, males are either rare or absent, and females may be winged or wingless and either lay eggs or give birth to living young.

In cold climates, some species have evolved very elaborate systems of alternation of generations (some with wings, some without), and on different food plants in response to day length, temperature and condition of the food plant. Often males are produced in autumn and are required to mate with females for production of over-wintering eggs. In Australia, the host alternation appears to be of little importance, and reproduction is mostly parthenogenic (females giving birth to wingless females without males present).

Many species of aphids are specific to one species of plant or to a few closely related plants, while some have a wide range of food plants. Most species develop into large colonies.

Many species live exposed on leaves, shoots or buds, and others live in concealment in folded or deformed leaves or in shelters constructed by ants. Many species are ant-tended, the ants avidly collecting the honeydew excreted by the aphids. Sooty mould is usually associated with aphids, sooty mould being several species of black fungi that grow on sugar.

Aphids are of great economic importance because of rapid development of large infestations and because many species are vectors of virus diseases. Some species have toxic saliva, which may cause stunting and deformity of leaves, shoots and twigs.

Fortunately aphids are preyed on by birds, marsupials and other mammals, lizards, spiders, mites and predatory insects. In particular, they are the principal food of numerous ladybird beetles, hoverflies and lacewings, and are parasitised by minute wasps. They are also subject to fungal diseases.

Aphids can be controlled with systemic insecticides.

Superfamily Coccoidea, families Coccidae, Diaspididae, Margarodidae, Eriococcidae and Kerriidae (scale insects) [Illustrations 29 to 37, 137, 164, 206 to 211] There are many different types of scale insects, both introduced and native species. Some native species are little known and associated with native plants only. There are species with cottony, felted, mealy or waxy coverings, naked or soft scales and others with leathery or hard coverings. Cochineal is made from one Mexican species, and in Asia two of the wax scale species are farmed in a small commercial way for wax production, as is one of the lac scales for production of shellac.

Several scales (exotic and native) attack a wide range of plants including fruit trees and ornamental plants. Some induce the growth of a great deal of sooty mould on the host or nearby plants. The saliva of others is toxic to plants and causes dieback of branches and death of plants; consequently scale insects are of great economic importance.

Scale insects are small to minute in size and feed by sucking sap from the host plant. Females may give birth to living young or, more often, eggs are laid either beneath the body or scale of the female or in a wax ovisac.

The first-stage nymphs (called crawlers) have functional legs and move freely on the host plant. The later nymph stages of a few species still retain the power of movement, but in the majority of species, the later nymphs and adult females are sedentary, the legs having been either reduced or lost and the long very fine stylet or mouthpart remaining inserted in the plant.

The adult female scale is usually a wingless enlarged version of the nymph. Males are unknown in some species, and in others may be wingless or two-winged, without mouthparts, and are very tiny, delicate, short-lived insects.

Scale insects are eaten by birds, marsupials, bugs, ladybird beetles, hoverflies, lacewings and the larvae of some moths. In addition, scale insects are parasitised by minute wasps and by several species of fungi (entomogenous fungi). [Illustrations 329, 330, 331.]

Chemical control of most species can be achieved by spraying with systemic insecticides or horticultural oil products.

Superfamily Coccoidea, family Pseudococcidae (mealy bugs) (These comprise the true mealy bugs and there are some mealy bug-like species in family Margarodidae) [Illustrations 38 to 40, 179, 238, 239] Many species of true mealy bugs live on native plants and some, which are pests of crops, are a problem to farmers and nursery workers.

In general, the true mealy bugs live more or less in concealment in folded leaves or close around growing points or on the roots of plants. Frequently they are ant-tended and the presence of ants betrays the presence of mealy bugs.

Eggs are laid, usually in a waxen ovisac attached to the female, or deposited on the surface of the food plant. The nymphs usually cluster together in a concealed and protected part of the plant. They feed by inserting their very fine mouthparts into the plant tissue and sucking sap. Their feeding causes mottling and deformity of leaves, and they can transmit virus diseases.

The adult females remain wingless and are typical mealy bugs in appearance, but the males are either unknown or tiny, delicate, short-lived winged insects.

The mealy bug-like species in the family Margarodidae include the solitary mealy bugs, some of which are very large and may reach 35 mm in length – but are harmless.

The predators of mealy bugs include birds, marsupials, spiders and in particular ladybird beetles, the caterpillars of some moths, hoverflies and lacewings. They are parasitised by tiny wasps and fungi.

For chemical control, use systemic insecticides.

Order Hemiptera, suborder Heteroptera, many superfamilies (bugs) [Illustrations 41 to 52, 173 to 176, 227, 228, 288]

Bugs make up a specific group of minute to large insects equipped with piercing and sucking mouthparts. Some bugs are somewhat shield-shaped, some very flattened for living under loose bark and others very slender. They make up a group of very diverse insects. Some species may be found living on all parts of plants; others are essentially terrestrial, living on the ground or in leaf litter; and still others are aquatic.

A few bugs are blood-suckers, living on birds and mammals including bats. The bedbug attacks man. Other bugs such as the so-called water scorpion or fish-killer feed on fish and other aquatic animals.

There are many predatory bugs [Illustrations 173 to 176, 228, 288] that feed on other small animals, including other insects and eggs of insects. Although they are beneficial when they feed on pests, they are not selective feeders and kill beneficial and harmless insects as well.

Plant-feeding bugs [Illustrations 41 to 52, 227] feed on leaves, shoots, fruit and seeds.

Many species give off a strong, offensive odour when disturbed and some can eject a caustic fluid that may be quite dangerous to the eyes.

The saliva of some bugs is toxic to plants and causes areas of dead tissues out of all proportion to the size of the bug or to the amount of feeding that has taken place [Illustrations 44, 45]. Several native bugs (citrus bugs and fruit-spotting bugs) feed on crop plants and are serious problems.

The adult female fastens the eggs, often arranged in a distinctive pattern, to plants. In some species (harlequin bugs), the female guards the eggs and newly hatched nymphs. The several nymph stages are like small, wingless adults.

Chemicals are less effective against adult bugs and are better directed at the nymph stages. Systemic insecticides are available for control, but when few bugs are present, control may be achieved by knocking the insects into a jam tin containing 2 or 3 cm of water and a little kerosene.

Order Thysanoptera (thrips) [Illustrations 60 to 64, 266]

Thrips are slender or elongate insects, mostly very small to minute in size. Fringed wings are characteristic of thrips ('thrips' is both singular and plural). Some feed on blossoms, fruit and leaves, and some cause galls or deformities on leaves and shoots. Other species feed on leaf litter and fungi, yet others are predators on other small insects (including thrips) and mites.

The natural enemies of thrips include birds, frogs, spiders, hoverflies and other predatory and parasitic insects and mites.

A few introduced species are serious pests of cultivated plants and their importance is increased by the fact that some are vectors of serious plant virus diseases.

Their mouthparts are such that they cannot pierce and suck in the normal way, but instead rasp or lacerate the surface tissues and suck or lap up the juices. Thrips damage is distinguishable as severe scarring.

Eggs are laid on the food plant and the several sizes of young thrips as well as adults live and feed together.

Chemical control can be achieved with systemic insecticides.

Order Neuroptera (lacewings) [Illustrations 181 to 186, 251]

Probably the best known of the lacewings are the larvae known commonly as ant lions, which construct funnel-shaped holes in dry, bare soil to trap ants and other small insects. Insects that fall into one of these holes are seized and sucked dry by the ant lion, which lies concealed in the dust at the bottom of the hole. The ant lion when fully grown pupates in the soil or attached to a nearby object, and later emerges as an adult, commonly known as a lacewing because of the lacy pattern of the wings. Adults of the soil-dwelling larvae generally deposit eggs singly in sandy soil.

The larvae of lesser known species live a free, active, predacious life on plants. Pupae of these species are generally attached to the bark or stems of plants. The beautifully winged adults are also predacious. To avoid cannibalism, the eggs may be elevated on long stalks and attached to plants singly or in clusters, sometimes arranged in patterns distinctive to the particular species.

Plant-dwelling lacewings are extremely beneficial, as both adults and larvae feed on pests such as psyllids, scales, aphids, mites, small caterpillars and other small insects.

In turn, lacewings are eaten by birds, bats, frogs, spiders and beetles and parasitised by mites and wasps.

Order Coleoptera (beetles) [Illustrations 8, 86 to 101, 111, 112, 161 to 166, 216 to 220, 241 to 244, 252, 253]

Beetles are by far the largest group of animals in the world. It has been estimated that approximately one-third of the known species of animals are beetles. They include some of the largest as well as some of the smallest insects.

The majority of beetles and their larvae feed on decaying vegetable matter and fungi or prey on other insects, some of which are serious pests of cultivated plants. The beneficial feeding habits of some beetles have been utilised by man; ladybird beetles (Coccinellidae), most of which feed on aphids, scale insects, mealy bugs and so on, have been used successfully to control pests; other beetles have been used to control noxious weeds; still others, such as the dung beetles which bury animal excrement, are used to control the bush fly. Some beetles play an important part in the recycling of nutrients by breaking down and burying plant and animal remains.

Many beetles and/or their larvae are serious pests, attacking timber, plants, seeds, grains and a whole range of stored products – food, clothing, textiles and leather.

Beetle damage to plants includes destruction of flowers by pollen feeders, chewing of leaves, ringbarking of branches, boring of branches and stems, chewing of roots and mining of leaves. There are aquatic beetles and the firefly is a beetle.

Adult females lay eggs on or in or near the food plant, and the larvae (grubs) feed by chewing.

The natural enemies of beetles include birds, marsupials and other mammals, lizards, spiders, wasps and other predatory insects. The eggs and larvae are parasitised by tiny wasps, mites and worms and are also subject to fungal, bacterial and virus diseases.

Chemical control of the adults often has disappointing results and the larval stages of borers are difficult to reach. Boring larvae may be controlled by pruning.

Order Diptera (flies, mosquitoes, midges) [Illustrations 1, 6, 116 to 121, 139 to 142, 286, 287, 328]

This very diverse order includes: mosquitoes; midges; sandflies; gnats;

fungus flies; hoverflies; houseflies; bushflies; blowflies; botflies; robber flies; crane flies; fruitflies; leaf-mining, stem-boring and gall-forming species; parasitic species such as the tachinid flies; and the blood-sucking tabanid or March or horseflies. Glow worms are the larvae of a fly.

Mosquitoes and flies outrank all other insects in medical and veterinary importance because they are vectors of many serious diseases. Some species of fly attack plants; in particular, fruitflies and leaf-mining, stem-boring and gall-forming species are of major economic importance.

On the other hand, flies are important pollinators of flowers and some plant-feeding species control weeds. Predatory species such as robber flies and crane flies may help to reduce the pest insects they feed on. Likewise, the parasitic species such as tachinid flies and hoverflies are very beneficial. Other species parasitise snails and spiders.

Mosquitoes have been used to spread virus (myxomatosis) in rabbits, and much of our basic scientific understanding of cytogenetics has been gained from study of the humble vinegar fly (often wrongly called fruitfly).

Adult flies generally deposit eggs in a moist situation or on food plants. The larvae or maggots are elongate and either have no legs or in a few instances have false legs.

Many breed on decaying vegetable or animal material. Fungus gnats, the adults of which are pests of greenhouse plants, breed in residues on the greenhouse floor.

The natural enemies of flies include birds, bats, reptiles, frogs, spiders and predatory insects such as dragonflies, wasps, other flies, beetles and bugs. They are parasitised by tiny wasps and flies, mites, worms, protozoa, algae, fungi and viruses.

Bait products are available for fruit fly control but their success has been variable.

Order Lepidoptera (moths, butterflies and their caterpillars) [Illustrations 65 to 81, 109, 114, 115, 122 to 136, 221 to 226, 245, 283]

This very large and diverse group includes some of the tiniest and some of the largest and most beautiful insects.

The group consists mainly of plant-feeders, but a few are carnivorous, feeding on the egg masses of other moths and butterflies, and on spiders, ant larvae or scale insects.

Some feed on stored food products and textiles (particularly wool), and a few species are ectoparasites on leafhoppers.

Almost all of the adults, the moths and butterflies are harmless to plants or may even be beneficial pollinators of flowers. Many are beautiful and the farming of butterflies in recent years highlights the interest in these insects.

Among the few harmful adults are the fruit-sucking moths of tropical and subtropical Australia.

Most of the larvae or caterpillars of moths and butterflies chew plants, and all parts of plants are attacked by one species or another. There are leaf-tying and leaf-eating species; blossom and fruit chewers; leaf miners and bark miners; branch and twig webbers; root feeders and borers.

Some live exposed on the surface of plants (some elaborately camouflaged), and others sew leaves together for shelter or tunnel in buds, flowers or fruit. Some construct a web-cover camouflaged with their own droppings, some shelter in communal bags or nests and some construct individual portable bags. Procession caterpillars can cause serious medical problems.

The caterpillars of some species are relatively uncommon and occur only singly or in very small numbers on plants. Such species are usually relatively harmless and the few leaves they eat are a small price to pay for the privilege of having an interesting caterpillar and a beautiful butterfly or moth in your garden.

Other species may occur in large numbers and can be very damaging to plants if not removed.

Eggs are laid on the food plant and, after several instars or moults, the caterpillars pupate and eventually the adult insects emerge. In general, moths are night flying and butterflies day flying, but there are many exceptions.

The natural enemies of moths and butterflies include tiny wasps that parasitise the eggs; the larvae are food for birds, mammals, reptiles, frogs, spiders, mites, bugs and wasps. Both larvae and pupae, as well as being subject to bacterial and virus diseases, are parasitised by nematode worms and a range of wasps and flies.

The adult moths and butterflies are preyed upon by birds, lizards, spiders, mantids, bugs, dragonflies, robber flies and other predacious insects, and are parasitised by mites and nematodes.

When necessary, caterpillars can be controlled with insecticides, particularly in the early stages.

Order Hymenoptera (sawflies, wasps, bees, ants)
This large group of insects includes many very highly specialised species. A vast number are predators and parasites (some complex hyperparasites). Others, such as some of the bees and ants, live in dense colonies and have highly developed social orders.

Members of the Hymenoptera have chewing and sucking mouthparts and the adults have a marked constriction between the abdomen and the thorax. Some of the groups of wasps, bees and ants have a venomous sting. A few of the Hymenoptera are miners and gall formers or plant-tissue feeders, while some feed on pollen and nectar. Most are important predators or parasites of other small animals.

Superfamily Tenthredinoidea (sawflies)
[Illustrations 82 to 85] Sawflies are sometimes called thick-waisted wasps because the adults generally lack the constriction between thorax and abdomen that is so conspicuous in most wasps.

The adult females of most sawflies insert their eggs into leaves or twigs by use of a saw-like ovipositor which slits the tissues. The caterpillar-like larvae feed on leaves. Other species with tunnelling larvae have an ovipositor adapted for boring. There are species with leaf-mining larvae and species whose larvae tunnel in soft stems or wood (for example, the introduced sirex wasp). The larvae of the tunnelling and mining species are either legless or have reduced legs.

A few species are predators of other insects and some may be parasites of wood-boring beetle larvae or may possibly feed on the fungus-infected frass in the tunnels of the beetle larvae.

Several introduced sawflies are established in Australia and include the sirex wasp, the cherry slug and a small leaf miner of pigweed.

The beneficial effects of the few predacious and parasitic sawflies seem to be outweighed by the damage caused by the plant-feeding species. Leaf-eating sawfly larvae can defoliate eucalypts, callistemons and callitris, and leaf-mining species can severely disfigure eucalypts and pongamia (now *Millettia pinnata*). Some species mass at the base of eucalypt trees and are eaten by sheep, causing death to the sheep.

Sawflies are eaten by mammals, reptiles and frogs and are parasitised by wasps and flies. In addition, they may be attacked by fungal, bacterial and virus diseases.

Sawfly larvae may be controlled by spraying insecticides, but it is practical to remove clusters of larvae manually.

Many superfamilies and families (wasps)
[Illustrations 3 to 5, 110, 143 to 153, 278 to 282] Wasps make up a large and highly

complex group of parasites and predatory insects. However, a number are gall formers and a few are miners, and some of these are pests of economic importance. Overall, this is a highly beneficial group of insects which exerts a very important control over many pest and potential pest species.

Some of the parasitic species are minute, others quite large. Some are primary parasites of insects, spiders and other small animals; others are secondary parasites or hyperparasites. Many parasitise eggs, others larvae, even larvae concealed in soil, beneath bark or boring in wood. Some parasitise pupae, and still others are either external or internal parasites of adult insects, spiders and so on.

Some place their eggs directly on or in the host, others near the host, or some wasp larvae actively seek out the host. Still others place their eggs in the food of the host and these do not develop until swallowed by the host. There are many intricate variations to the life cycles of the many parasitic species.

Many of the larger wasps actively seek out caterpillars, other insects and spiders, which they paralyse and on which they deposit eggs. Some excavate burrows and some construct mud cells or locate a suitable cavity which they provision with paralysed victims as food for their larvae.

The paper wasps construct papery combs similar to the wax combs made by some bees, and the maggot-like larvae in the cells are fed masticated caterpillars collected and prepared by the adults.

Many of the adult wasps are venomous stingers. However, they feed mostly on nectar and some are important pollinators of flowers.

Superfamily Apoidea, higher taxon Apiformes (bees) [Illustrations 187 to 192] Australia has a very distinct group of native bees. However, best known is the introduced honey bee. Bees are all highly beneficial insects, some being used for honey

production and pollination of crop plants. In fact, their value as pollinators of forest trees, fruit trees, pastures and crop plants is greater than their value as honey producers.

Some bees have a highly organised social caste system and feed their larvae specially prepared food throughout their development. In addition, there are many solitary species of bees which construct nests or burrows in soil, wood, pithy stems or in embankments of earth or rock.

Some bees are parasites on other bees, and the leaf-cutting bees are sometimes minor garden pests that disfigure plants by cutting neat, semicircular pieces from leaves to construct cells in the burrows in soil or holes in wood, using these cells as nests.

Some bees have stings and venom, but some native species such as *Tetragonula* do not. Some use wax in constructing nests and others use tree resin and mud.

Besides being attacked by predators such as mammals, reptiles, frogs, introduced toads, spiders, bugs and mantids, bees are also parasitised by beetles, mites and worms, and are subject to fungal, bacterial, virus and protozoan diseases. Introduced species of a beetle and a mite are causing serious problems for the future of bee-keeping.

Superfamily Vespoidea, family Formicidae (ants) [Illustrations 128, 137, 138, 236, 237] There are about 1000 species of Australian ants recognised at present and many more yet to be studied. They are all social insects, living together in colonies with well-defined castes consisting of workers and soldiers and often with an individual queen.

The males are usually winged and the females either deciduously winged or wingless. In addition, the worker castes are wingless. The larvae are legless maggots.

Some species of ant may construct nests in soil, others in wood, under stones or other objects, or by sewing leaves together. Several species may even move indoors where they become household pests. The nocturnal sugar ant or carpenter ant can damage furniture or fittings by chewing wood.

Ants may be predators or may scavenge dead insects and food litter. Some harvest seed and many tend or farm insects that excrete honeydew. In this latter regard they carry plant pests such as scale insects and aphids from one plant to another and these can spread virus diseases.

Some ants are venomous (have a sting in their tail) while others can only bite with their mouthparts.

Should action have to be taken against ants, it is better to pour a little dilute insecticide directly down into the nest rather than use a general area spray. Early in the morning after rain or hosing, the ants clean out their nests, which will then be clearly defined.

Chapter 3

PLANT DAMAGE – BY MOUTHPARTS, EGG-LAYING, BITES AND STINGS

The way in which pests damage plants depends on the type of mouthparts they have or how they feed. Some can bite and chew pieces of plant tissue while others can only pierce the surface and suck the juice or sap from plants. In between these two methods of feeding are several modifications such as rasping or cutting and sucking; rasping and swallowing; and chewing and sucking or lapping. Some pests combine their feeding with burrowing and tunnelling, mining, rolling or tying of leaves, or gall-forming.

A few adult insects damage plants during egg laying and not directly by feeding.

Piercing and sucking. Pests that feed by piercing and sucking include mites, the many members of the bug order Hemiptera (planthoppers, spittle bugs, cicadas, psyllids, lerps, white flies, aphids, mealy bugs and bugs) and also nematodes. Some nematodes feed on or in roots, others in leaves or flower buds. The adult of the fruit-sucking moth is also in this category. The moth pierces ripe fruit with its long haustellum (sucking tube) and sucks out the contents. The haustellum is tightly coiled when not in use.

Fortunately, fruit-sucking moths are confined to the tropics and subtropics, but the food plant for their larvae is becoming a common weed further south.

Rasping or lacerating and sucking. This is the method by which thrips feed. They rasp or lacerate the surface of young plant tissue and suck the juice that flows from the injury. Consequently their damage can be recognised as severe scarring.

Rasping and swallowing. Pest slugs and snails feed by licking or rasping away the surface of seedlings, soft young leaves and young roots.

Biting and chewing. Some of the animals in this category cause very dramatic plant damage such as complete defoliation. The most serious are grasshoppers (nymphs and adults), beetles (adults and some leaf-eating larvae), sawflies (larvae), caterpillars (larvae of moths and butterflies) and some of the phasmids or stick insects (nymphs and adults). Other biting and chewing insects include nymphs and adults of termites, crickets, cockroaches and earwigs. Occasionally a small amount of damage may also be done by springtails and slaters.

Chewing and sucking. Bees and wasps are equipped with chewing and sucking or lapping mouthparts. The leaf-cutter bee, for instance, cuts neat semicircles of leaf to construct the cells for its young. Bees and wasps suck up nectar from flowers and a few beetles are also able to take up liquid.

Borers and tunnellers. Among the most damaging insects that feed by boring and tunnelling are the termites or white ants. The larvae or grubs of some beetles also cause extensive damage by tunnelling or boring within stems, trunks or roots. Just occasionally, boring beetle larvae are found in wooden furniture.

The larvae or caterpillars of some moths also bore tunnels in stems and trunks. Some, such as the giant wood moth, feed and live entirely within the timber as borers before emerging as moths. Others may use a short tunnel (covered with webbing and their own droppings) for shelter, but feed on bark, sapwood or nearby leaves.

The larvae or maggots of fruitflies tunnel and feed within fruit and are a serious commercial pest.

Mining insects (leaf miners and bark miners). There are mining species in each of the following orders of insects: beetles (Coleoptera), sawflies and wasps (Hymenoptera), flies (Diptera) and moths (Lepidoptera). These insects are called miners because the tiny larvae mine or tunnel within the leaves or bark of plants. Usually the eggs are laid on the surface or are inserted in the tissue. The tiny larvae tunnel and feed below the surface and, in the case of leaf miners, within the leaf, making a characteristic scribbly pattern. Some species, particularly the larvae of beetles and sawflies, may make a larger blister or blotch mine.

Pupation may take place within the mine, or the larvae (particularly moth larvae) may leave the mine and pupate in the rolled edge or tip of the leaf. Some moth larvae mine leaves in the early stages, but later the caterpillars may live exposed on the leaf or in shelter made by rolling or joining leaves.

The larvae of leaf-mining beetles, sawflies, wasps and moths tend to be flattened in shape to suit the thickness of the leaf.

Some leaf miners severely damage crops (e.g. bean flies) and others severely disfigure ornamental plants such as azalea and some eucalypts. The characteristic scribbles on the bark of scribbly gums are made by mining moth larvae.

Leaf rollers and leaf tiers. Leaf rollers and leaf tiers include the caterpillars of some moths and butterflies, some thrips, ants, crickets and spiders. Caterpillars use the rolled or tied-together leaves as shelter and feed on the adjacent leaves or in some cases on the surface of the rolled leaves. The shelter is also used by the pupae.

Some species of thrips cause severe rolling or distortion of leaves, and all stages of the thrips shelter and feed together within the protection thus afforded. Thrips damage of this type could also be classified as galling.

Tree-dwelling crickets often join leaves together to construct a shelter in which they hide during daylight hours.

A few species of ants sew leaves together to make nests, sometimes, in the case of the tropical green tree ants, of considerable size.

Shoot- and twig-binders. Some spiders bind leaves or shoots together to provide shelter for their eggs and young. Although spiders do not feed on plants and are in fact beneficial because they capture moths and other flying insects, the binding can cause abnormal growth and becomes unsightly with trapped dead leaves and other debris. If necessary, bunches of dirty webbing, dead leaves and other debris can be removed by pruning or may be torn open by hand.

The larvae or caterpillars of some moths bind twigs and leafy branches of leptospermums, melaleucas and similar shrubs. The unsightly clumps of web, filled with chewed leaves and droppings in which the caterpillars shelter during daylight hours, can be removed by hand.

Gall-formers. There are gall-forming species in each of the following orders: flies (Diptera), wasps (Hymenoptera), psyllids, aphids, coccids, mealy bugs (Hemiptera), thrips (Thysanoptera), beetles (Coleoptera), moths (Lepidoptera), mites (Acarina) and nematodes (Phylum Nematoda). In addition, some species of rust fungus commonly form galls, particularly on acacias, and the bacterial disease crown gall can be a serious problem.

Some bacteria form galls on the roots of leguminous plants – peas, acacias and sennas. These *Rhizobium* bacteria are beneficial in that they fix nitrogen from the soil air for the benefit of these plants.

The harmful plant galls are mostly caused by species of flies, wasps, psyllids, aphids, coccids or mealy bugs. The gall is produced wholly by the plant in response to substances injected by the adult or secreted by the young developing pest. At the same time, eggs may be laid and, as the gall develops, the pests shelter and feed on the soft tissues or sap within the gall.

Some galls, particularly those affecting young stems, can be very damaging to plants and commercial crops. The citrus gall wasp is one of these. Leaf galls can be very disfiguring. The best treatment is early removal of galls and burning or wrapped disposal in a garbage bin to prevent hatchlings escaping.

Seed-feeders. The seed-feeding pests include beetles (adults and larvae), bugs (adults and nymphs), wasps (larvae) and the larvae of some moths. Some ants harvest seeds. Beetles (weevils) are serious pests of stored seeds. Some moth larvae feed on seeds either in storage or in green or ripening fruit on the plant. Bugs often feed by sucking from developing or ripening fruit and nuts and sometimes from freshly fallen seeds on the ground. Some species of tiny wasp lay eggs in maturing fruit or seeds and the larvae enter the seeds and feed on the contents. The adults may emerge from the empty seeds after they have fallen – or even in storage.

Reliable protection of seeds in storage is not easy for home gardeners, with the professionals using combinations of temperature control, sealed containers and insecticides to combat pests, with increasing levels of pesticide resistance.

Damage caused by egg laying. This type of damage is not associated with feeding and it includes the zigzag egg-laying cuts made on twigs by cicadas, the ringbarking of branches by some longicorn beetles and the puncturing of fruit by fruitflies.

Cicadas and fruitflies damage plants with their ovipositors. The ringbarking beetles chew neat rings of bark from small branches as part of the egg-laying procedure of cincturing branches to increase the starch (or food value) for their developing grubs. You may be aware if a wire is tied tightly around a branch it gradually swells on the upper side because of the accumulated starch from photosynthesis by the leaves.

HAVE YOU BEEN BITTEN OR STUNG?

Logically, biting would be carried out by the mouth and stinging by a defensive organ similar to a hypodermic needle carrying venom, in the rear end. However, while some creatures have mandibles or jaws for biting, many others have piercing and sucking mouthparts only, but are nevertheless regarded as biting (an example is mosquitoes).

The main confusion concerns ants, some of which have mandibles only (mouthparts for biting, such as meat ants), while others (jumper ants, lawn greenhead ants and fire ants) have pinchers on the front end as well as a venomous stinger in the rear. The introduced fire ant holds on with its jaws and rotates its body, inflicting a ring of painful stings. The arboreal green tree ant of the tropics adds a variation. It has only mandibles for biting – no stinger – but spits or dribbles citric acid into the bite, creating a sensation similar to a sting.

Potential biters include centipedes, ticks, spiders, termites, mantids, grasshoppers, crickets, predatory bugs (these have a very painful 'bite' because of their potent saliva), bedbugs, stink bugs (such as the citrus bugs, which can eject a caustic fluid into eyes), beetles (adults and grubs – some have formidable mandibles which can inflict a painful bite), mosquitoes, sand flies or midges, tabanid flies (March or horseflies), some ants, lizards and snakes. The danger of being bitten by venomous spiders and snakes is well documented and medical help should be sought immediately.

Potential stingers include scorpions, earwigs, wasps, some bees and some ants. The effect of some hairy caterpillars is simply mild irritation but others give a severe sting and even cause severe allergic reaction.

Some ants and some termites have the ability to direct a minute spray or jet of chemical or gas into the face of predators – nasute termites from the head, the ants from the rear.

Other creatures with interesting defence mechanisms include the bombardier beetles, which have a special chamber beside the anus and into which they can eject a couple of chemicals; these react with an audible explosion or pop and produce a minute cloud of caustic gas, which is ejected into the face of a pursuing predator. Another species ejects a sticky irritant substance into the face of predators.

The ultimate defence mechanisms are possibly the electric charges or shocks produced by some sea creatures.

The sharp leaf tips of some plants (such as macrozamia) and thorns of others apparently bear a toxin that causes wounds to temporally ache and glands to swell.

Chapter 4

PLANT DISEASES AND USEFUL, HARMLESS AND BENEFICIAL ORGANISMS

Plant disease may be defined as 'any disturbance of normal life processes, other than attack by insects and other small animals, resulting in abnormal growth, temporary or permanent check to growth, or premature death of part or all of a plant'. Accurate identification of diseases is important because they affect plants in different ways; they spread by different means and control methods are quite different depending on the disease. Identification can be tricky because the symptoms of plant diseases are frequently confused with those of pests and physiological disorders. Refer to Chapter 2 for the classification system.

Diseases are broadly classified into two categories according to their cause:

1   parasitic disease caused by parasitic bacteria, algae, fungi, viruses or seed-bearing plants such as mistletoe and dodder (nematode problems are often classified as diseases)

2   non-parasitic diseases or physiological disorders resulting from unfavourable growing conditions such as extremes

of weather or soil conditions, or from horticultural mismanagement.

BACTERIA

Bacteria are very small and can be seen only under a microscope. They are simple, free-living, single-celled organisms, usually rod-shaped, spherical or spiral. Some form clusters or chains. They were once classified as plants, but are now classified as a separate group, prokaryotes in the kingdom Monera. They reproduce simply by dividing in two. Some bacteria are able to produce nutrients by photosynthesis, while others have to obtain all of their nutritional requirements from their environment.

There are both saprophytic and parasitic bacteria, the first being mostly beneficial in that they help decompose plant and animal litter and thus recycle nutrients. Some parasitic species are also beneficial because they parasitise pests of cultivated plants. Other bacteria play essential roles in soil chemistry and plant nutrition by facilitating availability of elements and compounds to plants, and having symbiotic (mutually beneficial) relationships with plants, where bacteria colonise and obtain nutrition from plants while assisting host plants to access particular nutrients.

Bacteria can multiply very rapidly into enormous numbers. For example, the bacteria that cause bacterial wilt in tomatoes enter the roots from infected wet soil and multiply in the fine water-conducting channels in the stems. Bacteria soon clog these channels and the infected plant wilts and dies from lack of water – even in wet soil. The same occurs with cut flowers in vases contaminated with slime bacteria.

Serious diseases of beans and tomatoes and soft rots of fruit and ornamental foliage plants are also caused by bacteria. Such diseases may be spread by planting infected seed. They may also be splashed onto plants from infected soil or spread from plant to plant, usually under wet conditions.

The best way to manage bacterial diseases is to avoid infection by using disease-free seed or planting material, having long crop rotation cycles and not planting in infected soil. Generally fungicides (with the exception of copper-based remedies) are ineffective against bacterial diseases. Antibiotics have been used but, besides being extremely expensive, their horticultural use is seriously ill-advised because it places in jeopardy their medicinal value for treating human diseases.

ALGAE

Algae photosynthesise as do green plants. They occur in most habitats and vary from small single-celled types to large multi-celled forms such as the giant kelps. Though once classed as plants, they are now generally placed in the kingdom Protista. Of interest to this publication are the very few species that occasionally cause dead spots on the leaves of a few plants. If algae infestations ever warrant treatment, a fungicide such as a copper-based preparation could be an option.

FUNGI

Although once regarded as plants, fungi are now usually placed in the kingdom Fungi. Some species are single-celled entities, most consisting of microscopic branched threads (mycoplasma), clusters of which may be visible to the naked eye. Some produce larger, spore-bearing structures such as mushrooms and toadstools.

Some species of fungi, particularly yeasts, are important in brewing, wine-making, bread-making and medicine. Several species are cultivated or harvested for food and some (mycorrhizae) are beneficial, living symbiotically on plant roots. There are many groups of fungi – all reproduce by spores, fragmentation or fission. Some spores are produced very simply by vegetative means, others by a more complex sexual process, and yet other fungi use vegetative spores for spreading during the growing season and sexual spores for surviving cold or dry weather.

Fungi lack chlorophyll and have to obtain nutrients from live or decomposing plants or animals. Some fungi are saprophytes and some parasites. A few actively capture nematodes. The saprophytic species are very important decomposers of organic matter, although a few can change from saprophytes to parasites and attack living plants. Some of the parasitic fungi are beneficial in that they parasitise caterpillars, beetle larvae, scale insects and other pest animals.

Many diseases of plants are caused by fungi. Common fungal diseases include leaf spots, leaf blights, fruit rots, flower blights, lesions or cankers of twigs and stems, stem rots, collar rots, bulb and rhizome rots, root rots and wilts.

Some groups of fungi, such as powdery mildew, live more or less externally on the host plant and send short feeding tubes into the plant tissue. Fungi that live internally in plant tissue may send out threads bearing the spores or produce minute spore-carrying bodies – sometimes cup-like – on or in the surface tissue.

A few fungi, such as smut of cereals, are so closely involved in the cell structure of the host plant that initial examination reveals only stunted growth. A few fungi produce hard fruiting bodies, called sclerotes, similar to hard seeds, which can survive in the ground or compost heap from one growing season to the next.

A wide range of fungicides is available for control of fungal diseases. Some of the modern fungicides are quite specific to a small group of fungi and have little effect on other fungi. Other fungicides have more general application. Fungicides, like all other chemicals registered for use on crop and garden plants, are subject to extensive testing for safety and effectiveness, but safety is assured only if used responsibly according to label instructions.

Registrations are constantly reviewed, so it is not possible to provide a list of products that will remain up-to-date. Always follow current product labels and check current registration status. Commonly used broad spectrum products include those with active ingredients such as Mancozeb™, Mancozeb™ plus copper, copper sprays, sulphur (wettable, lime sulphur) and commercial mixtures, some of which contain more specialised products (such as rose spray containing synthetic pyrethroid, a systemic fungicide and possibly a product such as imidacloprid).

VIRUSES

Virus diseases are caused by infectious particles very much smaller than bacteria and can be seen only under an electron microscope. They are different in structure from the accepted ideas of living material and they are absolute parasites that can multiply only in the living cells of other entities. Bacteria and fungi can be grown on culture media in a laboratory and studied – but viruses cannot. A virus particle enters a plant cell and reorganises that cell to produce more virus particles as well as plant cells. Until availability of the electron microscope, diagnosis of plant viral infections relied on the transfer of suspected diseased plant sap to known indicator plants that could be relied on to grow recognised symptoms of each of the known virus diseases.

The most common symptoms of virus infection in plants are: mottling; concentric ring patterns; distortion of leaves, flowers or shoots; or breaks or streaks in the colour of flower petals. Infected plants are usually stunted and plants degenerate over a long period. Once a plant is infected, all parts of that plant will be carrying virus particles and there is virtually no cure – unlike with animals, which have an immune system and which may overcome a viral infection.

Virus diseases are spread in infected sap, a few quite simply by brushing against infected and then healthy plants. Outbreaks of the devastating tomato mosaic disease were found to be from the fingers of workers whose hands were infected from tobacco from infected tobacco plants. Other virus diseases are transmitted by vectors such as sap-sucking aphids and leafhoppers or the rasping and sucking thrips. Nematode transmission has also been recorded. Of the virus diseases transmitted by vectors, transmission of some is limited to a particular species, while others may be transmitted by several or many species. Virus diseases may be transmitted on secateurs and a few only by grafting. Few viruses are transmitted through seed.

Valuable plants in exceptional circumstances have been grown under constant high temperature for some time and then, if the plant survives, propagated by tissue culture from the growing tip to obtain virus-free plants from an infected plant. These are highly specialised procedures and in practice virus infected plants should quickly be removed and burnt to prevent spread of the disease.

As scientific knowledge advances, the causal organisms for some recognised diseases are found to be different from the original determinations. Thus some diseases previously determined to be virus infections are now found to be due to more recently discovered entities such as viroids and prions. They appear to affect plants and to be transmitted in the same way as viruses.

LICHEN

Many gardeners and professionals still question the effect of lichen on trees and shrubs.

Lichens are complex organisms composed of fungus and either algae or cyanobacteria combined in an intimate symbiotic relationship. They thrive just as well on fence posts, timber poles and rocks as on trees. Those that grow as epiphytes on trees are not parasites and do not take anything from the tree on which they grow. However, a thick covering of lichen may reduce photosynthesis by reducing available light. Lichens can be eliminated by spraying yearly with a copper spray.

Chapter 5

DEALING WITH PESTS AND DISEASES

Pests have potentially high reproductive rates because they have to survive attack from many parasites and predators, as well as seasonal variations. They also have to be able to take quick advantage of seasonally available food plants and favourable changes in the weather.

Some gardeners are prepared to accept a higher level of damage to their ornamentals, fruit or vegetables than are others, especially if the produce is for immediate home use. Commercial growers, on the other hand, have to harvest products early to ensure they are sound enough to survive long periods of transport and storage. Home gardeners have the advantage of being able to enjoy the improved flavour of fully mature, plant-ripened products.

Pesticides should be applied only if they are really necessary. Plants have evolved with some pests and there is evidence that suggests that small numbers of pests may even stimulate plant growth.

No plant or garden is better for having a pesticide applied if there is no need for it. Be prepared for the unexpected. Sometimes insecticides can damage the plants you are trying to protect. These phytotoxicity symptoms include leaf-fall, distortion or scorching of young leaves. Frequent and heavy applications of pesticides, especially where quite unnecessary mixtures are applied, seriously interfere with the natural balance of pests, parasites and predators. Interference with parasites and predators can lead to rapid increases in the populations of pests they normally keep in check.

In artificial garden situations the natural biological control of pests is not as effective as it is in natural, undisturbed bush environments. Also, garden hybrids and cultivars bred for quick growth and succulent, tender texture tend to be more attractive to pests than naturally occurring varieties.

CORRECT USE OF PESTICIDES

For your own safety and the sake of your garden and the environment, always follow the instructions on pesticide labels. The better your understanding of the pest and the pesticide you plan to use, the better the results. The basic principle is to apply the lowest effective recommended concentration of the pesticide to the most susceptible stage in the life cycle of the pest. Pests are not uniformly susceptible to pesticides throughout their life cycles. The larval or nymphal stage is generally when control measures are best applied, and the presence of adults can mean it is too late for the best effective treatment.

Because pests are increasingly developing resistance to pesticides, pesticide labels now indicate pesticide groups. If you spray regularly, you can minimise contributing to pesticide resistance by avoiding the regular use of products from one pesticide group.

Where overlapping generations of a pest occur – that is, young pests as well as eggs, as usually occurs with scale insects – two applications of pesticide may be necessary. The first will kill off the young crawlers present but will miss the eggs still sheltered under the adult females' scales. They require a second spray about 14 days later.

There is usually a delay of about 12 to 24 hours between pesticide application and death of the pest. Furthermore, some pests such as scale insects may stay attached to plants for days or weeks even though they are dead. Believing that they are still alive often leads to further unnecessary applications of pesticide. Some gardeners imagine that pest or disease damaged leaves should become normal again after pesticide treatment. While the damage is still obvious they go on spraying and often complain about the ineffectiveness of the pesticide. A test of whether pests are still alive is to crush several to see if they are dry (dead) or moist (alive).

Some of the chemicals used by commercial growers are, for safety reasons, not available in home garden packs, and are not registered for use in home gardens. Check www.apvma.gov.au for the latest safety guidelines and current registration status for home garden use of chemicals.

As with human and pet medications, be aware of what you have in storage and its age and expiry date. It may be useful to mark the purchase date on the container. Any out-of-date chemicals should be disposed of according to your local council guidelines.

MIXING CHEMICALS

Frequently, gardeners wish to mix several chemicals together to avoid the effort involved in spraying the garden several times. Some people add a foliage fertiliser as well. The surprising thing is that plants are not damaged more often by such mixtures. Always follow label directions regarding mixing of pesticides and other products. Incompatible mixtures may cause burning or other phytotoxic plant reactions, or they may result in one or all of the ingredients being rendered ineffective.

Several all-purpose mixtures are marketed. These preparations usually contain a fungicide, an insecticide and a miticide. They generally give satisfactory results while only small numbers of insects, mites or fungal spores are present, but should an outbreak of a particular pest or disease occur, it is advisable to use a remedy for that specific problem for one or two applications, instead of the all-purpose preparation.

TYPE OF SPRAYER

Most pesticides for home garden use are now available in pressure packs, which are applied to the plant from the container. Several types of pump-up sprays are available and appear satisfactory in home gardens where a constant pressure for exact delivery of spray is not vital.

Use separate sprayers and measuring equipment for weed-killers and pest-killers, or you may find tiny residues of weed-killer in pesticide are enough to damage a valued plant. It is virtually impossible to remove all traces of herbicide from spray equipment.

CARE WITH PESTICIDES

ENVIRONMENT CARE

Before spraying or dusting an apparent pest, find out what it is, what it is doing and whether it is harmful to your plant. If it is feeding on your plant, is it doing sufficient damage to warrant removal? Can it be simply rubbed or pulled off? If a chemical must be used, use the right chemical for the particular pest.

OPERATOR CARE

Virtually all the chemicals used to control pests are poisonous. They have to be poisonous to kill pests. The product label will provide instructions for safe application of pesticides. You will need the correct, well-maintained equipment and personal protection for the job, no matter how small.

Before opening containers of chemicals, read the label and instructions carefully. Know what chemical is in the product you are spraying – not just the brand name. Follow the directions carefully and measure the right amount. It is potentially dangerous to guess the quantity or put in extra. Use a metric measure rather than risking mistakes by converting back to imperial. The rates of dilution have been carefully worked out by chemists, and any extra chemical might damage either the plants or you.

The most dangerous material is the concentrate in the bottle. Beware the drip that runs down the outside of the container and may inadvertently get on your hand after you have removed your gloves.

Follow all safety directions on the container. Material safety data sheets are available online at www.apvma.gov.au.

PLANT CARE

Do not spray if the temperature is expected to be above 32°C (90°F). On a hot day, it is safer to spray early in the morning on turgid leaves rather than late in the afternoon when the leaves are limp. If rain is threatening you will be wasting chemical and your time if you proceed with spraying – check the weather forecast. Do not spray drought-affected plants. As a precaution in dry times, plants should be watered thoroughly a day or two before spraying.

Once pesticides have been mixed with water, they should be used within a few hours. Mixed sprays are likely to become ineffective in storage, with some chemicals losing effectiveness very quickly.

PESTS ASSOCIATED WITH FLOWERS, FRUIT, SEEDS, LEAVES AND SHOOTS

The majority of the insects, mites and other small animals that are recorded on garden plants are to be found at some time or other amongst the leaves or on flowers, fruit or young shoots. Accordingly, the illustrations are grouped as much as possible in this way, starting with pest or damaging species of animals in Illustrations section I and moving to the less harmful, harmless and beneficial animals in Illustrations section II, and so on.

The plant-feeders range from species that chew to species that suck sap. A few are attracted to the nectar and/or pollen. Some of these, in particular some beetles, may cause considerable damage by tearing apart the flowers in the scramble for pollen.

Young leaves are the most popular target for plant-feeding pests and consequently a large number of photos illustrate species that feed in one way or another on leaves.

The animals that damage young shoots are often difficult to determine because frequently they move away after feeding. In other instances, the pest is either concealed within the damaged tip or is so small or so well camouflaged as to make detection difficult.

For further information relating to the subjects of the following illustrations and insight into the relationships between the various creatures depicted, the reader should refer to the appropriate section in Chapters 2, 3 and 5, and Appendix I.

GALL FORMERS
See Chapters 2 and 3, also Illustrations 180, 213, 214, 247, 302, 318, 332, 333, 336

1 Xerochrysum gall fly

Procecidochares utilis

(a) This insect was introduced from Mexico and released in the hope that it would control Crofton weed –
Ageratina adenophora. Unfortunately it readily attacks the native *Xerochrysum* species, as this galled stem illustrates.

(b) An opened gall showing the tiny white spherical larva.

(c) An opened gall showing the pupa or chrysalis.

(d) The adult fly.

Adult length 6 mm

Family Tephritidae

2 Brachychiton leaf gall
These unusual, bottle-shaped galls are quite common on the underside of the leaves of *Brachychiton discolor*. Heavily infested leaves are seriously disfigured. The galls are caused by the larvae of an unidentified insect.

3 Lophostemon flower and leaf gall
(a) The galls formed commonly on the buds and leaves of *Lophostemon suaveolens* are often brightly coloured. They result from the development of the larvae of a tiny wasp that feed and pupate within the galls.
(b) The top has been lifted off this gall to expose the pupa.
(c) The wasp normally emerges from the pupa within the gall some time before the top splits off the gall and allows the tiny wasp to escape. These galls are rarely a serious problem. If necessary, early removal by pruning and burning or wrapped garbage disposal will reduce infestations.
Adult length 2 mm
Order Hymenoptera

4(a)

5(a)

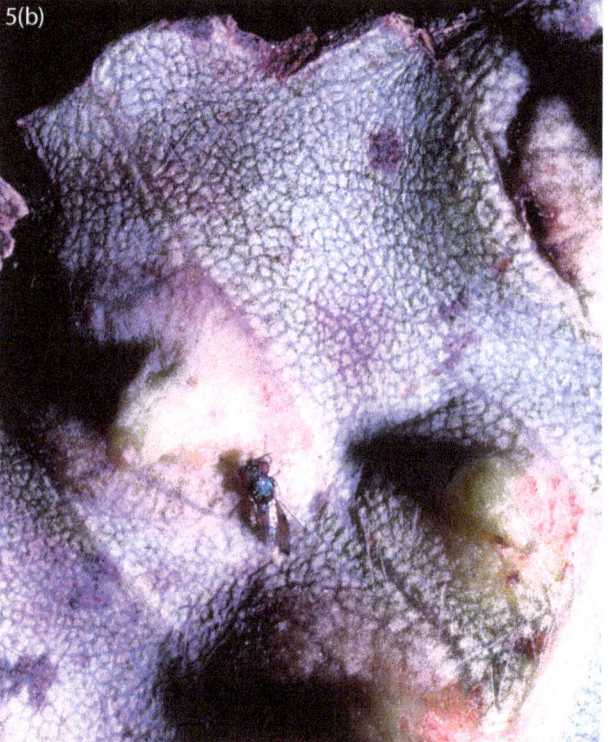

5(b)

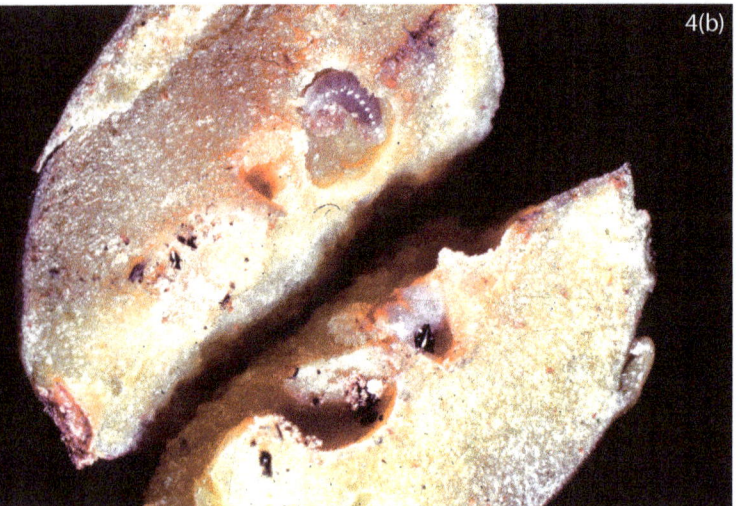

4(b)

4 Acacia flower gall

(a) There are several different causes of galls on acacias. (See Illustration 302.) The two lower spikes of buds illustrated on this acacia have been attacked by a tiny gall-forming wasp. Eggs have been laid in the spikes, resulting in the formation of the large fleshy galls.

(b) This cross-section shows one of the individual cells and the wasp larva or maggot. Pupation takes place within the gall and on emergence the adult wasp chews its way out of the gall. If necessary, early removal of galls by pruning then burning or wrapping and placing in a bin will reduce infestations.

Adult length 2 mm

Order Hymenoptera

5 Lophostemon prickly leaf gall

(a) These hard-pointed galls occur on the leaves of *Lophostemon confertus*. A heavy infestation is quite disfiguring.

(b) The galls are caused by the larvae of this tiny wasp. If considered necessary, galled leaves may be removed and disposed of to limit infestation.

Adult length 0.5 mm

Order Hymenoptera

6(a)

6(c)

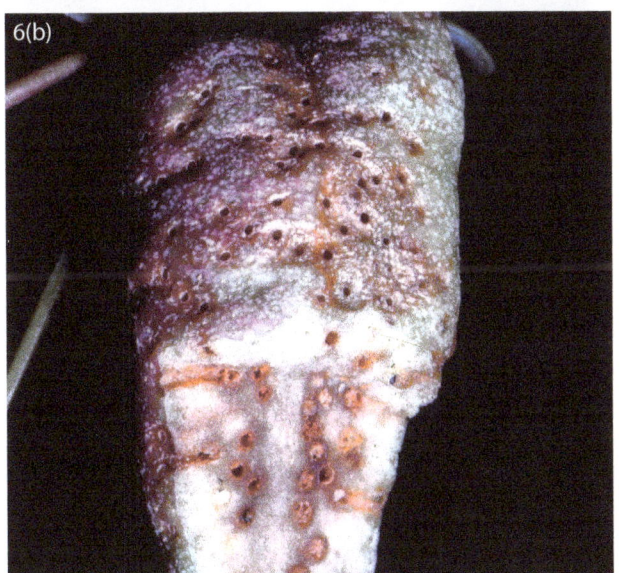

6(b)

7

6 Eucalyptus nematode or fly gall

(a) These remarkable galls result from infestation by a microscopic nematode worm that is carried by a tiny fly. The nematode larvae are deposited in the leaves at the same time as the fly lays its eggs. The fertilised female nematodes enter the bodies of the female fly larvae and are carried through the fly pupal stage into the adult female flies. Galls may be formed on young shoots, on leaves, or on young stems.

(b) Many emergence holes can be seen where adult flies have emerged from the gall, while the cut-away section shows many separate cells within the gall.

(c) These *Fergusonina* sp. flies have emerged from galls. If galls are picked off and burnt or wrapped for bin disposal, infestation may be reduced.

Fly: *Fergusonina* sp. Adult fly length 2 mm. Family Fergusoninidae

Nematode: *Fergusobia* sp. Family Neotylenchidae

7 Eucalyptus bubble gall

Glycaspis sp.

Sometimes eucalypt leaves are heavily infested with the soft bubble-like galls caused by this insect. If necessary, infested leaves could be pruned off and disposed of. Family Psyllidae

8(a)

8(b)

9

9 Leptospermum gall
These very colourful galls are caused by an unidentified insect that attacks leptospermums in some districts.

8 Eucalyptus tip gall
(a) These peculiar fleshy galled tips appear to be formed by fusion of the young leaves.
(b) This small weevil has emerged, its larva or grub apparently having either caused the gall formation or opportunistically used the gall tissue as food. If necessary, galls could be removed and burnt or wrap-disposed.
Adult length 4 mm
Family Curculionidae

10

10 Eucalyptus stem and leaf gall
Apiomorpha pileata
These unusual galls are frequently covered with mealy, white, waxy secretions of the causal insect, which is closely related to mealy bugs. The stem galls are caused by females, the leaf galls by males.
Family Eriococcidae

11 Banksia mite gall
These grotesque galls on banksia fruit result from an
infestation of microscopic mites that feed and shelter
amongst the deformed tissue. Early removal and burning or
wrapped disposal of deformed fruit will reduce infestation.
Microscopic size
Family Eriophyidae

PSYLLIDS
See Chapters 2 and 3, also Illustration 7

12 Grevillea bud drop psyllid
(a) *Grevillea rosmarinifolia* and similar species suffer bud drop caused by this minute psyllid. Sparse flower spikes may be a familiar sight and infestation may also cause leaf curl of young shoots. A close examination with a hand lens will usually reveal the presence of the minute nymphs, which cause buds to turn yellow and drop off.
(b) The tiny speckled-winged adult (on the lower left leaf) may shelter among the buds or foliage. Spraying may be necessary.
Adult length 1.5 mm
Family Psyllidae

13 Grevillea flower psyllid
(a) Flower spikes of some grevilleas, such as *G. hodgei* and *G. juncifolia*, may be severely blighted by this tiny psyllid. The minute nymphs as well as adults live amongst the buds and feed by sucking sap.
(b) The white, waxy secretions of the nymphs are the obvious sign of their presence. The young shoots may also be attacked and distorted. Severe infestations may necessitate spraying.
Adult length 1.5 mm
Family Psyllidae

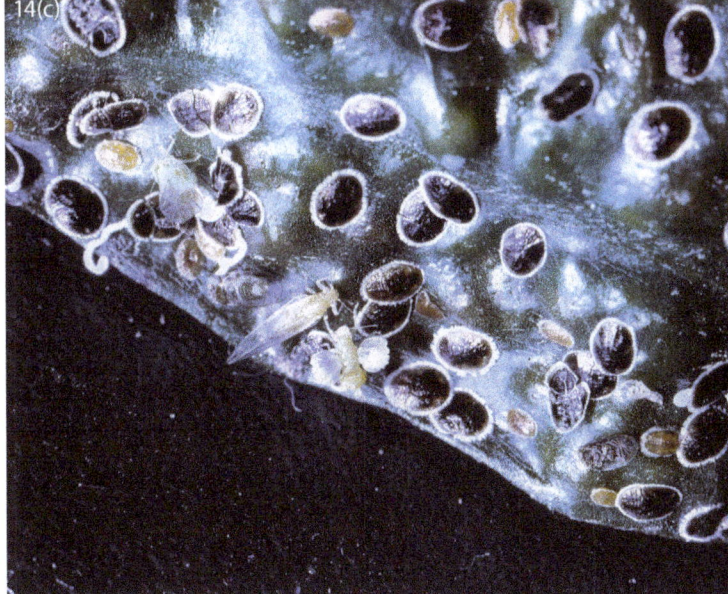

14 Pimple psyllid
Trioza eugeniae

(a) The pimple-like distortions on the young growth are caused by the nymphs of a tiny psyllid on syzygiums, callistemons and eucalypts. Sometimes young stems are also distorted.

(b) The countless first-stage nymphs initially move freely on young growth and attract predators such as ladybird beetles.

(c) Later-stage nymphs settle down and become scale-like and embedded, causing characteristic distortion, before hatching into winged adults. White wax is secreted by the nymphs.

Adult length 2 mm

Family Triozidae

16 Hibiscus woolly psyllid
Mesohomotoma hibisci
Each nymph is densely covered with long, white, waxy filaments, and as the nymphs live closely together in large numbers, infested leaves and shoots have a white, woolly appearance and are sometimes mistaken for mealy bugs. Chemical control may be necessary.
Adult length 2 mm
Family Carsidaridae

15 Banksia leaf-curl psyllid
(a) Severely distorted old growth was caused by a tiny insect when the growth was young.
(b) Severely distorted young growth with insects present
(c) The tiny last-stage nymph – length 1.5 mm
Family Psyllidae

17 Bunch psyllid on grevillea

(a) The very severe effect of the feeding of this minute insect is caused by its toxic saliva. It is illustrated here on *Grevillea sericea*, but also recorded on *Acacia fimbriata*.

(b) The severely deformed leaves and shoots are tightly bunched together and shelter the hard-to-see tiny orange insects.

(c) Two of the tiny insects can be seen. Chemical control is necessary.

Nymph length 0.5 mm

Family Psyllidae

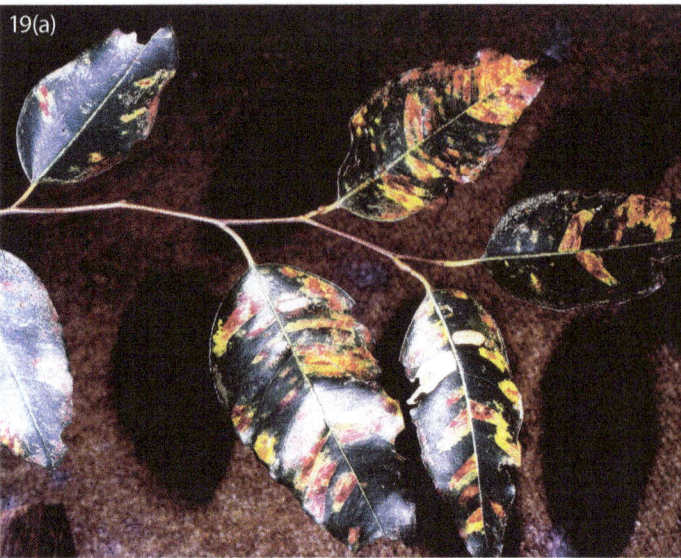

19 Lace lerp
Cardiaspina sp.
(a) Psyllids that form protective covers are sometimes called lerps. They pierce and suck from leaves and their toxic saliva causes leaf discolouration.
(b) The feather-like, intricate covering or lerp is quite beautiful and shelters the nymph stages. An adult is also shown. It is best not to plant trees that are susceptible to such psyllids.
Adult length 3 mm
Family Psyllidae

18 Melaleuca and baeckea psyllid
Ctenarytaina sp.
(a) Curling or twisting of young growth may indicate the presence of this small, inconspicuous, sap-sucking insect. Some white webbing may be present.
(b) The tiny yellow adult is shown. Chemical control may be necessary.
Adult length 1.5 mm
Family Psyllidae

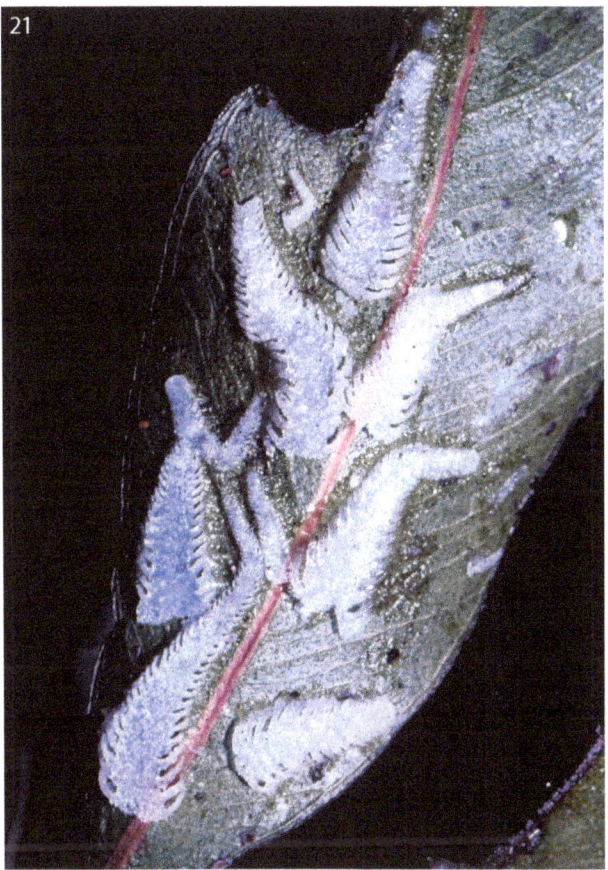

20 Shell lerp
Hyalinaspis sp.
This species of psyllid constructs a delicate, shell-like lerp. Its habits are more or less identical to the preceding species.
Adult length 3 mm
Family Psyllidae

21 Lerps
Anoeconeossa sp.
These are the intricate and beautiful coverings or lerps of another species of psyllid that causes sooty mould because of the copious honeydew they excrete. Some lerps have a sugary taste.
Adult length 2 mm
Family Psyllidae

APHIDS
See Chapters 2 and 3, also Illustration 240

22 Aphid colonies
Aphis nerii
(a) Aphids vary in colour from green, yellow, brown to black depending on species. They multiply very rapidly into large colonies. Their toxic saliva distorts young growth and they carry virus diseases. Chemical control is necessary. This yellow species infests *Asclepias* sp. and related plants.
Adult size 2 mm
Family Aphididae
(b) A dark-coloured species of aphid on *Atriplex* sp.
Adult size 2 mm
Family Aphididae

LEAFHOPPERS
See Chapters 2 and 3, also Illustrations 177, 178, 203 to 205

23 Leafhopper flattid
Colgar peracutum
(a) The pointed-nosed, rocket-shaped nymphs cover the leaf surface with waxy powder. They jump if disturbed.
(b) Although the green sail-shaped adults and nymphs suck sap, they do not appear to be important pests.
Adult length 9 mm
Family Flatidae

24 Jassid or leafhopper
(a) The tiny jumping insects wander about on the underside of leaves, sucking sap and leaving a faint whitish line visible on the upper side of the leaf.
(b) Both nymphs and adults live and feed together.
Adult length 1–3 mm
Family Cicadellidae

WHITE FLY
See Chapters 2 and 3

25 White fly
Orchamoplatus mammaeferus
The adults and eggs of a common pest species of white fly are shown. Their eggs are arranged in a pattern which is characteristic of this species.
Adult length 2 mm
Family Aleyrodidae

26 White fly
An unidentified scale-like species of white fly.
Adult length 2 mm
Family Aleyrodidae

27 Orchid white fly
This orchid pest is generally regarded as a scale insect, but is in fact a white fly. Note the ant seeking honeydew.
Adult length 2 mm
Family Aleyrodidae

28 Callistemon white fly
After an active early stage, the nymphs of this species soon settle down to a stationary, scale-like existence. The damage they do to plants and the method of control are similar to those of scale insects.
Adult length 3 mm
Family Aleyrodidae

SCALE INSECTS
See Chapters 2 and 3, also Illustrations 137, 164, 206 to 211

29(a)

29(b)

29 Chain scale
Pulvinaria sp.
(a) Most species of scale insects are stationary on plants after the first-stage nymphs, but this scale keeps constantly on the move. The tiny scales, which hatch from the egg sacs on the leaves, move down the plants and then back to the leaves while feeding and developing. As the scales move one behind the other, they resemble a green chain. The white insect shown is a predator – *Cryptolaemus* sp. (See Illustrations 164, 208.)
(b) As the mature scales move back to the leaves, the females turn into cottony egg sacs. This scale causes the growth of a great deal of sooty mould.
Adult female length 2 mm
Family Coccidae

30

30 Large flat green scale
This insect is one of several poorly known native scales. It gives off a sugary solution, attracts ants and causes sooty mould.
Adult scale length 10–12 mm
Family Coccidae

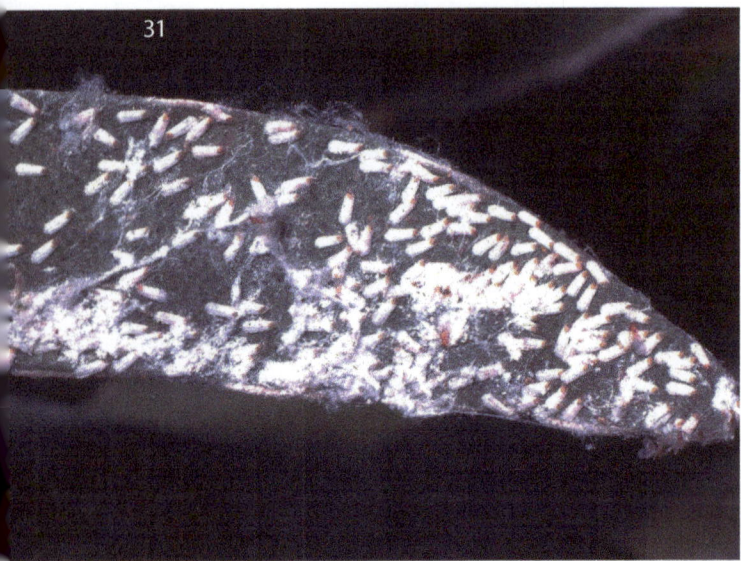

31 A mussel scale
This unidentified scale has been recorded on melaleucas and callistemons but, as with several other native scale insects, little is known about it at present. Note the winged males.
Adult female length 1.7 mm
Family Diaspididae

32 Circular black scale
Chrysomphalus aonidum
This scale has a toxic saliva and infestations can cause dieback of leav and branches.
Size 2–3 mm
Family Diaspididae

33 Pink wax scale
Ceroplastes rubens
(a) The waxy covering protects the insect from many predators and also from some pest-killing sprays. The eggs hatch under the female scale and, after a short period of moving around, the young scales settle down in a fixed position. The tiny pink young crawlers can be seen here emerging from the upturned female and crawling on the leaf.
(b) This is a particularly dirty scale because it excretes large amounts of honeydew, and infested plants become covered with sooty mould.
Adult female length 4 mm
Family Coccidae

34 Plumed scale
This is another little-known scale insect that seems to restrict its feeding to the broad-leafed melaleucas. This scale is quite large in size (up to 8 mm) and is attractively plumed.
Adult female length 8 mm
Family Coccidae

35 Tessellated scale
Eucalymnatus tessellatus
This quite large, flat, brown scale appears to feed mainly on the leaves of lepidozamia and related plants. Notice also one pink wax scale and several small mussel scales.
Adult female length 5 mm
Family Coccidae

36 Fern scale
Pinnaspis aspidistriae
(a) Yellow spotting and dieback of fronds of staghorn, elkhorn and bird's nest ferns are better known than the tiny scale that causes the damage.
(b) This tiny scale amongst the brown sporangia is quite destructive and can cause the death of ferns in one season if it is not controlled.
Adult female length 1.5 mm
Family Diaspididae

37 Leptospermum scale
This is one of the white louse scales and infestations cause dieback of branches and even death of plants. It also causes some sooty mould.
Scale length 1 mm
Family Diaspididae

MEALY BUGS
See Chapters 2 and 3, also Illustrations 179, 238, 239

39 Common mealy bug
Pseudococcus sp.
This mealy bug attacks a range of plants and while it has little effect on the vitality of the plants its appearance and the associated sooty mould are highly objectionable.
Adult length 6 mm
Family Pseudococcidae

38 Bulb mealy bug
Bulbous plants may be attacked by mealy bugs, which work their way down the leaf bases to the bulb in the soil. The leaves on affected plants come up mottled and distorted, and the symptom resembles that of a virus disease. Clusters of mealy bugs of various sizes can often be seen if the leaf bases are pulled apart.
Adult length 4 mm
Family Pseudococcidae

40 Hibiscus mealy bug
Maconellicoccus hirsutus
This species commonly attacks hibiscuses and their relatives. It damages the growing points and causes right-angled bends in the growth. Note the predatory cryptolaemus larva (see Illustration 164).
Adult length 3–4 mm
Family Pseudococcidae

PLANT BUGS
See Chapters 2 and 3, also Illustration 227. See also predatory bugs

41(a)

41(c)

41(b)

41 Larger horned citrus bug
Biprorulus bibax
(a) This native bug has become a common orchard pest, causing fruit drop. All stages of nymphs and the adults pierce and suck from young shoots and young fruit, causing them to yellow and fall. Removal and destruction by knocking them into a jam tin partly filled with a little water and kerosene is often sufficient to reduce damage. However, care should be taken to avoid the caustic fluid that these bugs eject.
(b) The first-stage nymphs at first cluster together, but soon disperse and live individually.
(c) Eggs are laid in a geometric pattern.
Adult length 2 cm
Family Pentatomidae

42 Bronze orange bug
Musgraveia sulciventris
(a) This native insect normally feeds on the native citrus species, but has become a serious pest of commercial citrus. All stages of nymphs and the adults pierce and suck from young shoots and young fruit, causing them to yellow and fall. The caustic fluid ejected by this bug is dangerous to the eyes.
(b) The different stages of nymphs vary in colour, giving the impression that there are several species of insect present. The eggs are like tiny bubbles or spheres and are laid in a characteristic raft-like pattern.
Adult length 2.5 cm
Family Tessaratomidae

43 Fruit-spotting bug
Amblypelta nitida

(a) The adult fruit-spotting bug. There is also a very similar banana spotting bug, *Amblypelta lutescens*.

(b) These nymphs of the fruit-spotting bug are shy and difficult to detect, but can be distinguished by the unusual, thickened section of the antennae. All stages of this bug feed on a range of plants, including exotics and macadamias, and are common causes of fruit drop. This bug has also adapted to several commercial fruits and is a very serious pest.

(c) The final-stage nymph before turning into the winged adult.

Adult length 1.3 cm
Family Coreidae

44(a)

44(b)

44(c)

44 Acacia leaf-spotting bug
Rayieria tumidiceps
(a) These bugs have an extremely toxic saliva; where they pierce and suck from a leaf, a ring of damaged tissue quickly expands and becomes necrotic.
(b) Adult bugs on a newly attacked leaf.
(c) This was the fate of a sand-mining revegetation planting.
Adult length 1.5 cm
Family Miridae

45(a)

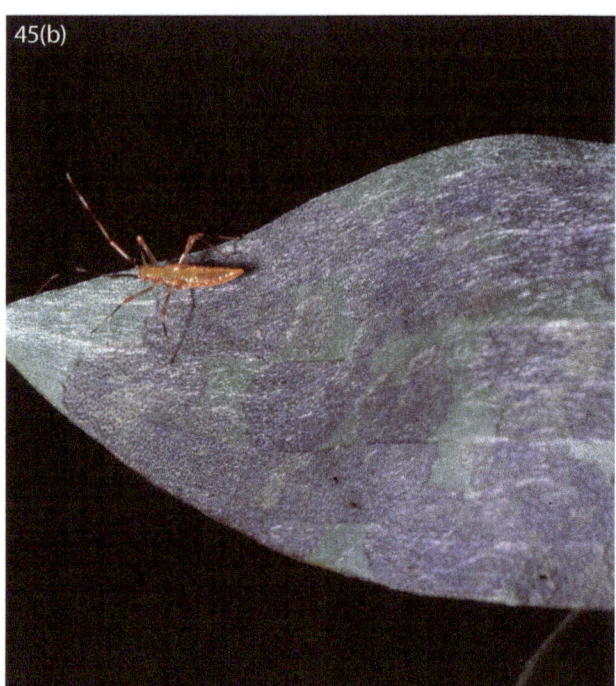

45(b)

45(c)

46

45 Melaleuca leaf-spotting bug
Eucerocoris suspectus
(a) The blighting of new shoots and young leaves by this delicate, shy bug is out of all proportion to the size of the insect and the number of bugs that can be found on severely affected plants. Its saliva is extremely toxic to plants as shown by the spots of dead tissue everywhere it has pierced the leaf surface. It feeds mainly on plants of the Myrtaceae family.
(b) The prettily marked, delicate, shy and hard-to-find nymphs also cause extensive spotting of young leaves.
(c) An adult bug resting on the expanding tissue damage from recent feeding sites. Because of the small number of bugs per tree and the difficulty finding them, chemicals are rarely applied for this bug.
Adult length 1 cm
Family Miridae

46 Callistemon tip bug
Pomponatius typicus
Soft growing tips of callistemons may become tufts of dead leaves with no apparent cause, and melaleucas may be similarly affected. This bug, which pierces and sucks the tips, is the culprit. It is the same colour as the dried tips and is very well camouflaged as it clings motionless among the dead leaves. Periodic collection and destruction by hand may give sufficient control.
Adult length 1.8 cm
Family Coreidae

47 Harlequin bug
Tectocoris diophthalmus
(a) Two species of bug are called 'harlequin' – this one and the less colourful *Dindymus versicolor*. This bug commonly feeds on native and exotic hibiscuses and related plants. There is a striking difference in colour between adult males and females, the male being the more colourful.
(b) The female stands guard over the eggs and for a short time over the newly hatched young.
(c) The colourful nymphs vary in colour with each moult.
Adult length 2 cm
Family Scutelleridae

48(a)

49(a)

48(b)

49(b)

49 Eucalyptus tip bug
Amorbus alternatus
(a) All stages of this common bug suck from young shoots of eucalypts and cause blighting of the young growth.
(b) In contrast to the drab adults, the nymphs are very brightly coloured. Periodic collection and destruction by hand may give sufficient control.
Adult length 2.2 cm
Family Coreidae

48 Aralia bug
Amorbus rubiginosus
(a) This large bug feeds on umbrella trees and other aralias, severely damaging the new leaves and growing points with its toxic saliva.
(b) The damaged tips are severely blighted.
Adult length 3 cm
Family Coreidae

51 Ehretia lace bug
Dictyla sp.
Lace bugs are usually very specific to particular plants and will not touch anything other than their particular food plant. This species feeds in ehretia trees. Note the winged adult and the spiny-looking nymphs.
Adult length 3 mm
Family Tingidae

50 Stephania lace bug
Stephanitis queenslandensis
(a) A leaf showing fairly typical bleached areas on the upper leaf surface from the feeding on the underside. The tiny bugs form colonies on the underside of leaves and their sap-sucking activities cause the damage.
(b) The adult bug is a beautiful, clear-winged insect with delicate vein patterns. Note the prickly, transparent and black-spotted nymphs. Lace-bug damage can be distinguished from mite damage because lace bugs dot and streak the underside of leaves with their excreta.
Adult length 2 mm
Family Tingidae

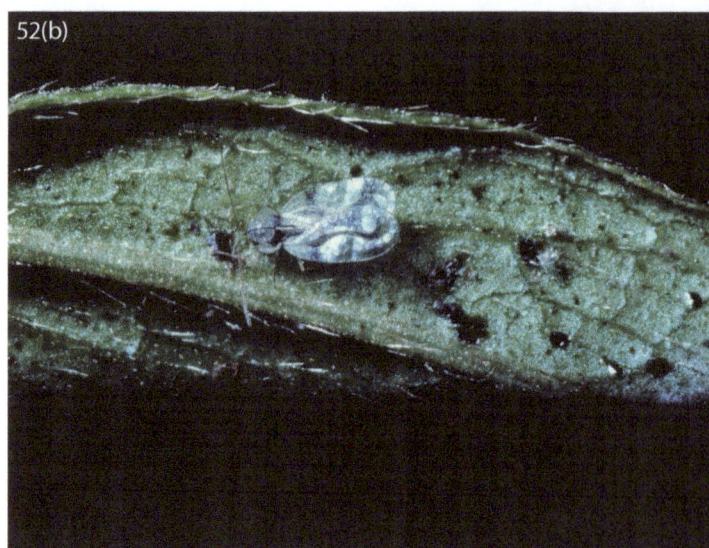

52 Azalea lace bug
Stephanitis pyrioides
(a) Feeding by this common azalea pest causes bleaching of the leaves, and an examination of the underside of leaves will reveal the typical spotting of excreta left behind by lace bugs.
(b) The adult bugs have beautiful wings.
Adult length 3–4 mm
Family Tingidae

MITES
See Chapters 2 and 3, also Illustrations 11, 195, 196, 212, 235, 263

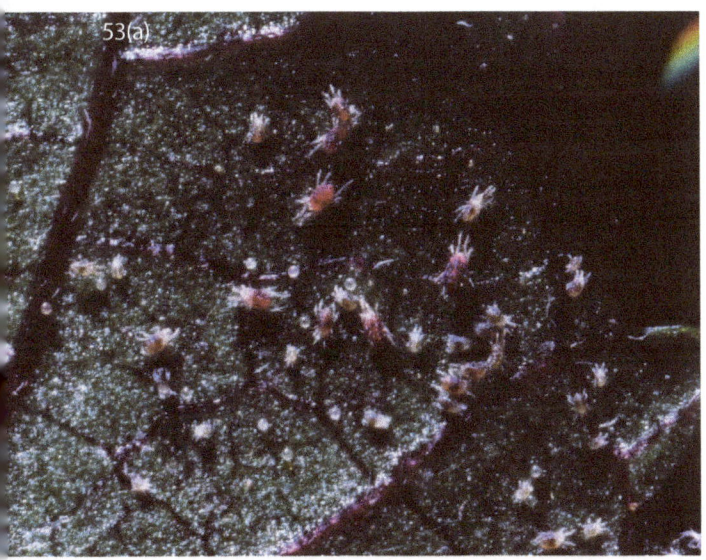

53 Red spider mite
Tetranychus sp.
(a) Red spider mites live and feed in colonies established mainly on the underside of leaves. They suck sap and cause bleaching or silvering of the leaves, which is visible on the upper leaf surface. Red spider mite damage may be confused with damage by lace bugs and thrips, but the mites do not mark the leaf with their excreta as do lace bugs and thrips.
(b) Greater magnification reveals the characteristic mite shape and eight legs. Spider mites may be red, yellow or more or less colourless, and sometimes with dark marks. The immature stages are usually paler than the adults and the eggs are microscopic, pale yellow spheres.
Adult length 0.7 mm
Family Tetranychidae

54 Brevipalpid mite (false spider mite, bunch mite)
Brevipalpus sp.
(a) Healthy foliage on left – mite-infested on right. The bunch mite feeds on upper and lower leaf surfaces and stems. It causes silver or slightly rusty discolouration, deformity of young shoots and sometimes leaf drop.
(b) Brevipalpid or bunch mites are smaller than spider mites and more brightly coloured, but their microscopic size makes them difficult to see. This illustration shows what you could expect to see through a ×20 hand lens.
Adult length 0.4 mm
Family Tenuipalpidae

55 Broad mite

Polyphagotarsonemus latus

(a) The broad mite feeds mainly on or within the folded growing point and on the very young leaves. The distortion that is characteristic of broad mite damage often becomes obvious after the mites have left the damaged parts. This is typical broad mite damage to seedlings.

(b) Typical broad mite damage to new growth, in this instance on *Tecomanthe* sp.

(c) Broad mites along with the previous species are sometimes called cyclic mites because they damage new growth but by the time the damage becomes obvious the mites have gone and only with difficulty can eggs be found. Note the severe rusty scarring of stem and leaves; such scarring would be followed by several centimetres of normal growth before the next infestation cycle.

(d) Broad mites are extremely small in size and more or less colourless. They are difficult for the inexperienced to see even with a hand lens. This photo shows what you could expect to see through a ×20 hand lens.

Adult size 0.3 mm

Family Tarsonemidae

56 Couch mite
Aceria cynodoniensis
This small, elongate mite attacks couch grass, causing deformed growth.
Adult length 0.2 mm
Family Eriophyidae

57 Blister mite
(a) Blister mite is a fairly common pest of *Corymbia ptychocarpa* and other eucalypts. The damaged areas of leaf may resemble blister-like raised sections when viewed from the upper surface, or sometimes small leaves may appear rough on both surfaces.
(b) The affected leaf surface is roughened and distorted and countless minute elongate mites can be seen.
(c) The extremely tiny and difficult-to-see mites are elongate or torpedo-shaped, and two pairs of legs are reduced or absent. This photo shows what you could expect to see through a hand lens.
Adult length 0.5 mm
Family Eriophyidae

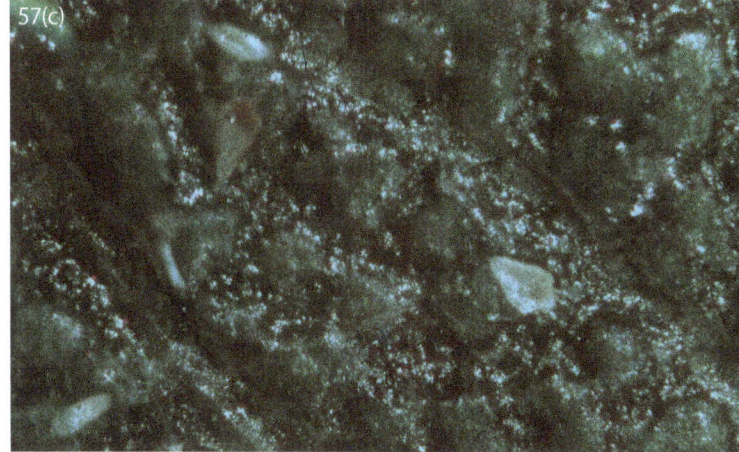

58 Erinose mite
The best known of the erinose mites is the species that severely deforms the leaves of lychee. Erinose mites are usually very tiny, often thread-like or torpedo-shaped and cause a felty surface, in which the mites shelter, to develop on the leaf surface.
Adult length 0.5 mm
Family Eriophyidae

59 Erinose mite
Erinose mites are usually specific to particular plants. This species deforms the young leaves of *Elaeocarpus eumundi*.
Adult length 0.5 mm
Family Eriophyidae

THRIPS
See Chapters 2 and 3, also Illustration 266

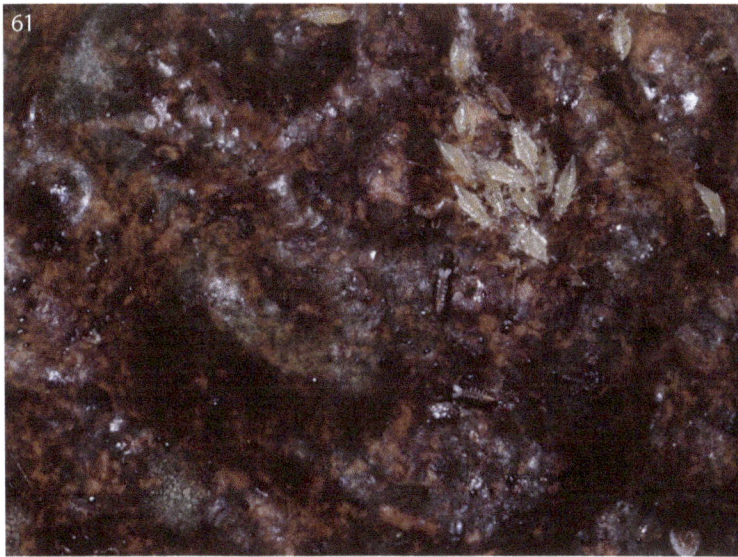

61 Leaf thrips
All stages of nymphs and adults live together and they dot the surface where they feed with their excreta in a similar manner to lace bugs. Their feeding causes severe, rough scarring, which is silver or brownish in colour. The nymphs are usually pale yellow or orange; the adults are very slender, dark-coloured or black insects and carry two pairs of tiny fringed wings.
Adult length 1.5 mm
Family Thripidae

60 Flower thrips
Infested flowers may develop brown streaks and may wither prematurely as a result of damage caused by the feeding of these minute insects. The tiny thrips are hard to see because they are so small and they hide in the florets or between petals or bracts.
Adult length 0.5 mm
Order Thysanoptera

62(a)

63(a)

62(b)

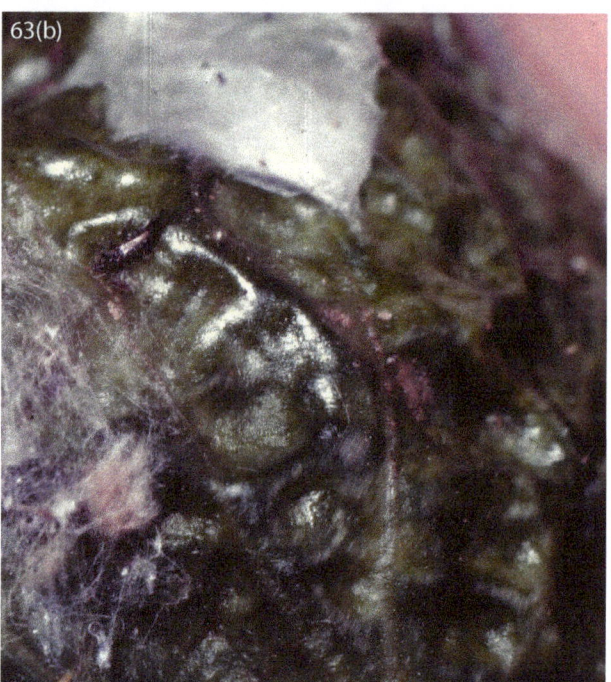

63(b)

62 Callistemon leaf-rolling thrips
Teuchothrips sp.

(a) Severe distortion of the leaves of callistemon is caused by these thrips. All stages of nymphs and the adults live within the tightly rolled leaves.

(b) A pale nymph and a black adult can be seen on the leaf tip. Control measures include pruning off affected shoots and, if necessary, spraying.
Adult length 2 mm
Family Phlaeothripidae

63 Piper thrips
(a) The gall-like rolled leaves caused by these thrips provide shelter in which the thrips multiply into very large numbers.
(b) Other creatures also take advantage of the shelter and, in addition to the black adult thrips, the photo shows spider eggs.
Adult length 2 mm
Family Phlaeothripidae

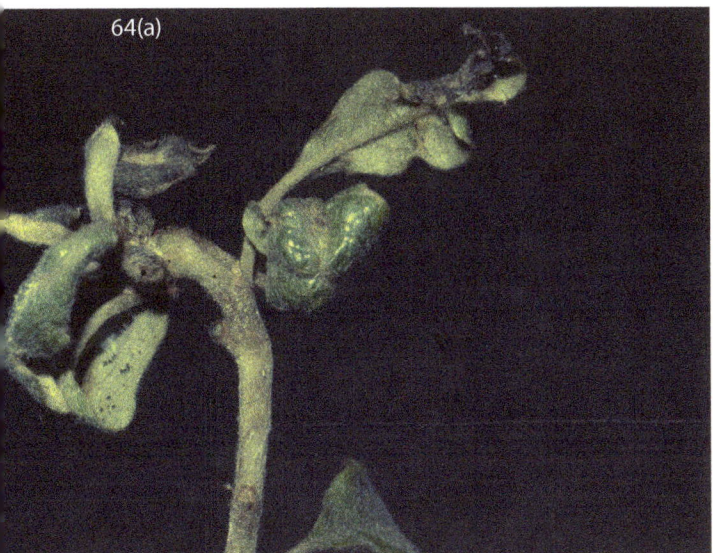

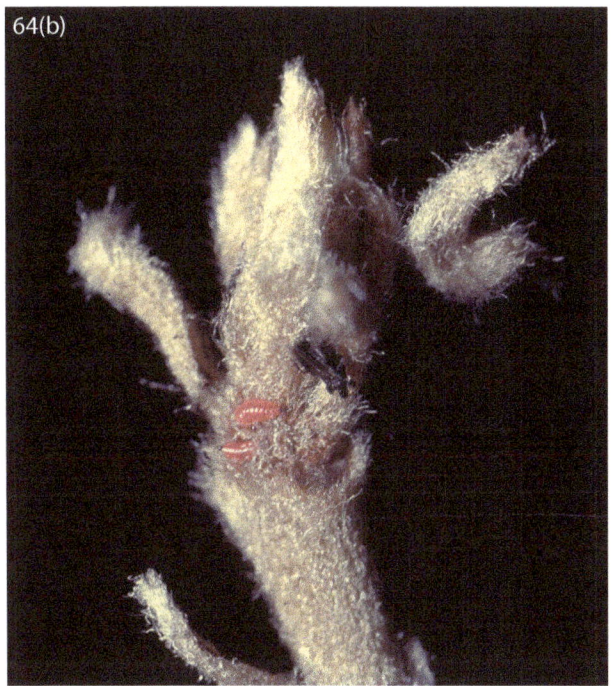

64 Pittosporum thrips
Teuchothrips ater
(a) Most species of pittosporum are susceptible to attacks by thrips. The feeding of this species causes sharp bending of new shoots and distortion of leaves. Other insects such as caterpillars often take advantage of the shelter offered by the deformed leaves.
(b) The pink or orange thrips, nymphs and the black adults shelter amongst the deformed leaves and stems and among the hairs on the plant surface.
Adult length 2 mm
Family Phlaeothripidae

MOTHS AND CATERPILLARS
See Chapters 2 and 3, also Illustrations 109, 114, 115, 122 to 136, 221 to 226, 245, 283

65(a)

Under Surface.

centre spot Top. Top wing varies.

scales small close shine

Upper Surface.

T.1.
Dec 15 1937 3 others 16/
Caught at Lamp
& Bananas.

65(b)

65 Fruit-sucking moth
Eudocima spp.
(a) Unlike other insect pests, the larvae are harmless and feed on vines such as *Stephania*. Several species of these large and beautiful moths attack a range of fruit and can cause major damage. Their long and very strong proboscis is coiled when not in use but can be extended to pierce the rind of fruit, enabling the moth to suck out the contents. The most effective means of control remain a torch and tennis racquet.
(b) This is typical of the larva, but may in fact be the larva of a closely related species. The food plant (*Stephania*) is becoming a widespread weed.
Wingspan 10 cm
Family Noctuidae

66 Elkhorn fern spore caterpillar

(a) This damage to elkhorn ferns, caused by a tiny caterpillar, is often mistaken for a fungal disease.

(b) The caterpillar tunnels and feeds within the brown spore pads, and it causes the death of the ends of the fronds.

(c) The tiny caterpillar is difficult to find and, if exposed, it quickly covers itself with spore cases.

(d) The fully fed caterpillars pupate under cover of the spore cases, and finally the adult moths emerge. Chemical control of infestations may be necessary to minimise damage.

Adult length 3 mm

Order Lepidoptera

67

67 Blossom caterpillar
This small, unidentified caterpillar is one of several different moth larvae that feed on or in buds and flower spikes. Some, such as the one shown, are hard to see because they tunnel in and around the buds.
Order Lepidoptera

68(b)

68(a)

68(c)

68 Orange palm dart
Cephrenes augiades
(a) Palm leaves may be very severely chewed by this caterpillar, which sews the edges of the leaflets together for protection while feeding.
(b) The caterpillars may pupate within the sewn palm leaflets or sometimes within the sewn leaves of nearby plants.
(c) The adults are called darts or skippers because of their unusual manner of flight.
Wingspan 3 cm
Family Hesperiidae

69 Grevillea looper
Oenochroma vinaria
This moth larva is a common pest of grevilleas and other proteaceae. One caterpillar can strip leaves off a branch.
Caterpillar length 6–7 cm
Family Geometridae

70 Hibiscus caterpillars
(a) Hibiscuses are very palatable to many insects and animals. Both of the caterpillars shown are the larvae of unidentified moths.
(b) Caterpillars on hibiscuses are voracious feeders and need to be removed either by hand or chemically.
Caterpillar length 4–5 cm
Order Lepidoptera

71(a)

71(b)

71(c)

71 Hawk moth
Psilogramma menephron
(a) These large caterpillars with a 'tail' grow to 10 cm length but are usually well camouflaged, and one or two on a healthy shrub could pass unnoticed. They are hungry feeders and, if present in larger numbers or on small plants, they may require removal or at least thinning out. Caterpillars could be picked off by hand rather than sprayed with chemicals.
(b) They usually pupate in the soil or amongst leaf litter on the soil.
(c) The large, handsome moth may have a wingspan of 9–10 cm.
Wingspan 9–10 cm
Family Sphingidae

72 A cluster caterpillar
These destructive, hairy caterpillars appear in large numbers. They live and move close together, usually just touching each other, and they feed mainly at night. Unless removed, they will soon defoliate plants. The adults are moths.
Order Lepidoptera

73 Tube caterpillar

Enchesphora lithochlora

(a) This species feeds mainly on the broad-leafed, paper-barked melaleucas. The caterpillars live in groups, webbing leaves together and feeding mainly on one surface of the leaves. Each constructs a tube-like covering of chewed leaf and excreta which, as the caterpillar grows, is extended back and forth through and around the leaves on which it feeds. The tube is enlarged in diameter as the caterpillar develops and, eventually, a sealed tip is constructed to hold the pupa.

(b) Part of the tube is shown on the left and extending below the moth to become the pupal shelter. It is rarely a serious pest but, if necessary, clusters of tubes can be pruned off and burnt.

Adult length 2.5 cm

Family Pyralidae

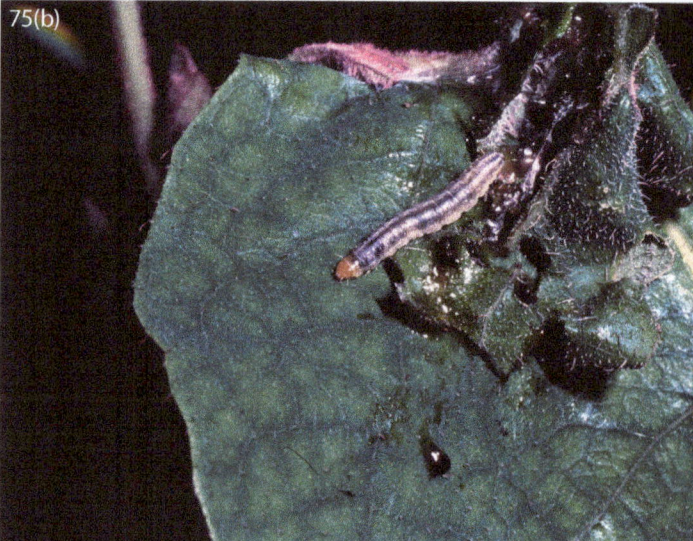

75 Leaf roller

(a) Most plants are at some time or other attacked by some type of leaf-rolling or tying caterpillar. Usually it is the young shoots that are tied together with silk webbing.
(b) The small caterpillar shelters, feeds and finally pupates within the protection of the tied leaves before the adult, a small moth, emerges.
Order Lepidoptera

74 Leaf and twig webber

(a) The caterpillars that cause this damage and disfigurement to leptospermums and sometimes melaleucas eventually turn into moths. They bind twigs and leaves together with a mass of webs filled with their own droppings and bits of chewed leaf.
(b) Several caterpillars live together within each shelter and at night they feed on nearby leaves. Pupation takes place within the web.
Family Pyralidae

76 Hibiscus leaf roller
Haritalodes derogata
This caterpillar cuts the broad leaves of hibiscuses across, but leaves enough attachment to usually keep the cut leaf alive. It rolls the cut section and shelters, feeds and pupates in the roll of leaf.
Family Crambidae

77(a)

77(b)

77(c)

77(d)

77 Callistemon tip borer

(a) This very persistent pest mainly attacks callistemons but sometimes also melaleucas. Light attacks probably do more good than harm because they are equivalent to tip pruning, but persistent attacks have a stunting effect and seriously distort the appearance of plants. The whole life cycle of this small moth is completed within 5–8 cm of the tip.

(b) The tiny larva or caterpillar tunnels and feeds down the centre of the young shoot.

(c) Pupation takes place within the hollowed-out shoot and the emerging moth cuts a small exit hole.

(d) The non-descript moths are tiny, quick-moving and hard to find.

Adult length 3 mm

Order Lepidoptera

78 Case moth
Hyalarcta huebneri
(a) This leafy case moth feeds on a wide range of plants
and is a serious pest because large numbers can breed up
quickly and host plants can be defoliated. They are best
removed by hand and crushed or burnt.
(b) A new hatch of this case moth showing how
prolifically they breed. As they develop they cover their
silken bags with pieces of dead leaf.
Family Psychidae

79(a)

79(b)

79(c)

79(d)

79 Bulb caterpillar
Brithys crini
(a) This is one of two moths/caterpillars (the other is *Spodoptera picta*) that seriously damage bulbous plants.
(b) The caterpillars feed on the leaves and flowers, tunnel or mine within the leaves and stems, and will also tunnel down into the bulb and destroy it.
(c) The nondescript dark moth has distinctive white hindwings.
(d) The excreta above the bulbs indicate the caterpillars are underground feeding on the bulbs.
Caterpillar length 40 mm, moth wingspan 40 mm
Family Noctuidae

80 Lawn caterpillar
Herpetogramma licarsisalis
These moth larvae (sometimes unfortunately called sod webworm) are night feeders, translucent, and hide in the turf during the day. Their presence is indicated by threadbare patches of lawn and flooding will bring caterpillars to the surface. Their presence is often indicated by the presence of a predatory wasp [Illustration 149]. Day-feeding caterpillars are unfortunately called army worm. ('Worm' should be reserved for the legless earthworms and nematodes.)
Adult moth size 2.5–3 cm
Family Crambidae

81 Azalea leaf-miner or roller
Caloptilia azaleella
The small caterpillar (moth larva) mines within azalea leaves in the earlier stages, but exits the mine and rolls the edge or tip of the leaf to make a shelter in which to pupate.
Adult moth size 8–12 mm
Family Gracillariidae

SAWFLY
See Chapters 2 and 3

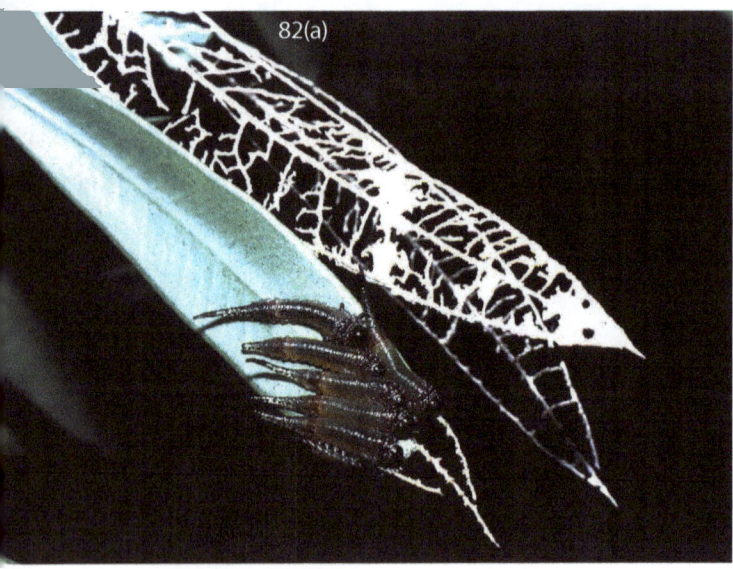

82(a)

82(b)

82 Callistemon sawfly
Pterygophorus insignis

(a) These very destructive caterpillar-like larvae quickly reduce leaves to mere skeletons. The larvae feed side by side, sometimes with a group of larvae on each side of the one leaf. When disturbed, they bend their heads and tails over their backs and they regurgitate to repel attackers. The fully fed larvae leave the food plant and wander singly for some distance to pupate.

(b) The wasp-like adult slits the edge of a leaf with its saw-like ovipositor and inserts the eggs. Freshly inserted eggs are shown beside the adult. This species is a serious pest and should be either removed by hand or sprayed.
Adult length 2.3 cm
Family Pergidae

83(a)

83(b)

83(c)

83 Sawfly larvae

Perga sp.

(a) This large species feeds on eucalypts and broad-leafed melaleucas. During the day, they cluster together on the stems but at dusk they move up onto the leaves to feed. When disturbed, they bend their bodies and regurgitate a fluid that smells of very strong eucalyptus oil.

(b) This similar species has similar habits. Sometimes the larvae congregate at the base of trees and are eaten by sheep, which subsequently die.

(c) The adult sawflies.

Adult length 2.8 cm

Family Pergidae

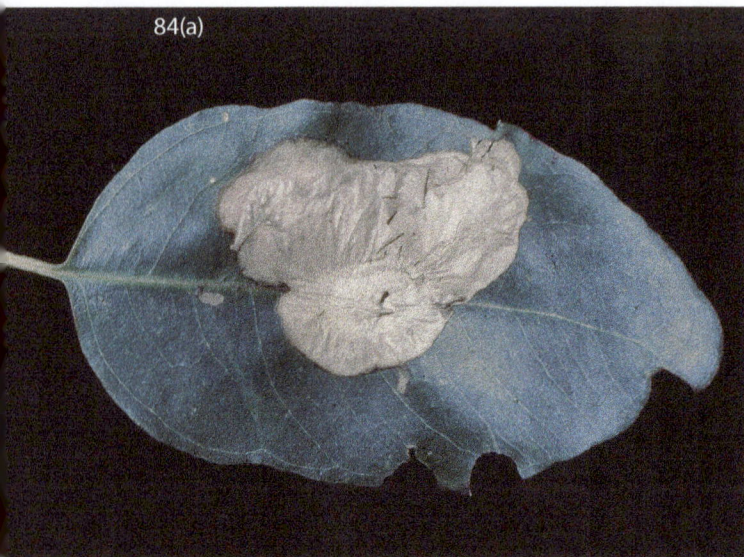

84(a)

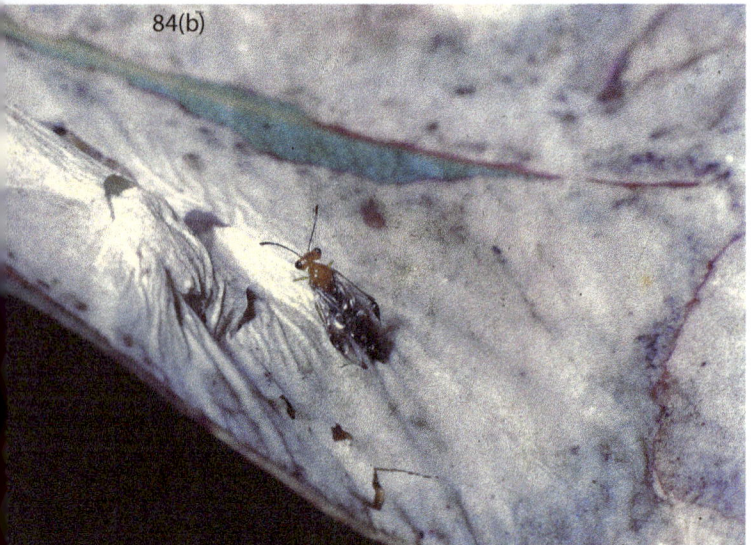

84(b)

84(c)

84 Blister sawfly
Phylacteophaga froggatti

(a) Some species of eucalypt and other plants such as *Pongamia* are seriously disfigured by this sawfly and at times almost every leaf may be mined.

(b) The pupa is formed within the mine and eventually the adult shown emerges. Frequently, affected trees are too tall to spray, but some gardeners treat young plants.

(c) Removal of the cuticle over the mined area reveals a flattened larva living among its own litter.

Adult length 8 mm

Family Pergidae

85

85 Blackberry sawfly
This sawfly attacks blackberry and some related plants.

Adult length 2.5 cm

Family Pergidae

BEETLES

See Chapters 2 and 3, also Illustrations 8, 111, 112, 161 to 166, 216 to 220, 241 to 244, 252, 253

86(a)

86(b)

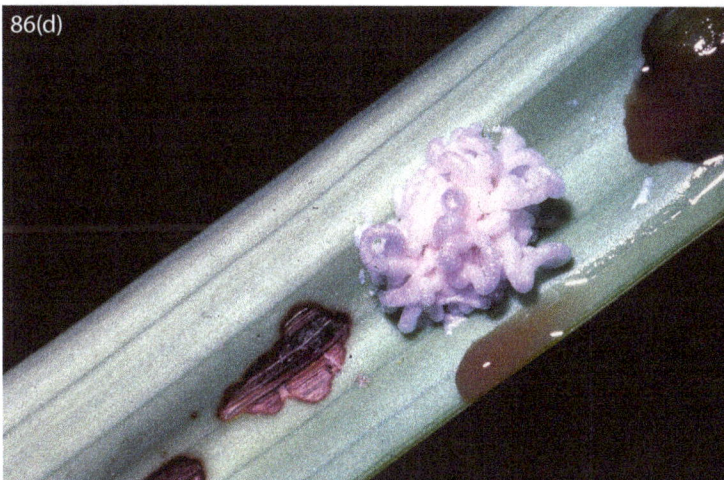

86(d)

86(c)

86 Orchid beetle
Stethopachys formosa
(a) This very destructive pest of orchids eats flowers, buds, leaves, new growth and, in this case, young seed pods. When approached, the adults usually drop to the ground, where they hide. They are hard to kill with chemicals, and collection night and morning is the best method of control. They may simply drop if you approach them with your hand or hat extended.
(b) The yellow eggs are attached to buds, flowers or new growth.
(c) Both the adult beetles and the maggot-like larvae or grubs are very hungry feeders. The larvae tunnel down and destroy new shoots.
(d) The pupae are covered with a white substance very like styrene foam. Sometimes pupae are attached to plants or more often they are partly buried amongst the roots.
Adult length 1.2 cm
Family Chrysomelidae

88 Blossom chaffer beetle
Protaetia fusca
This pollen-feeding beetle sometimes tears flowers apart while searching for pollen. The chaffer beetle usually occurs either singly or only in twos or threes and, therefore, is not a serious pest and the larva is unimportant.
Adult length 1.8 cm
Family Scarabaeidae

87 Nectar scarab or white clothes beetle
Phyllotocus apicalis
This common name is applied because the adults are attracted to anything white and frequently they swarm over washing hanging on the clothes line. The beetles are pollen-feeders, and they frequently swarm on and quickly destroy the blossoms on white-flowered plants. The larvae are unimportant and the adults are little affected by sprays.
Adult length 1 cm
Family Scarabaeidae

89 Monolepta or red-shouldered beetle
Monolepta australis
These very destructive beetles sometimes swarm into gardens or commercial orchards and strip the leaves, fruit skins or flowers of plants. Attacks may last a few hours or a few days before the swarm of beetles moves on. They breed in turf or pasture. Adult beetles are fairly resistant to chemicals and results from application of sprays are usually disappointing.
Adult length 6 mm
Family Chrysomelidae

90 Beetle larvae
Paropsis sp.
These beetle larvae are sometimes mistaken for sawfly
larvae because they feed together in groups on the leaves
of eucalypts. If necessary, groups of larvae can be picked
off by hand. Spraying is rarely necessary.
Larva length 8–12 mm
Family Chrysomelidae.

91 A leaf-eating ladybird
Paropsis sp.
These attractive beetles feed mainly on eucalypts and
occasionally they occur in large numbers and defoliate
young eucalypt trees in gardens.
Adult length 12 mm
Family Chrysomelidae

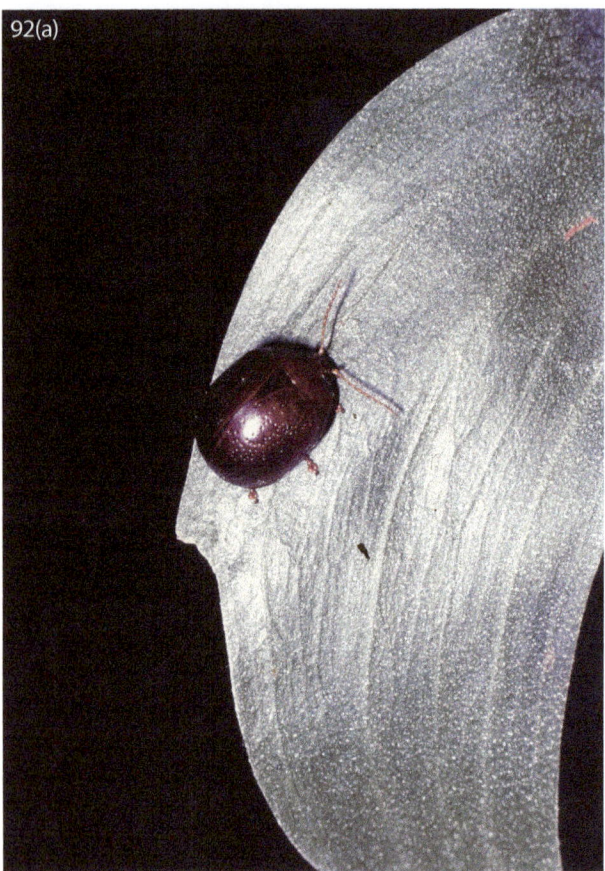

92(a)

92(b)

92(c)

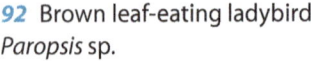

92 Brown leaf-eating ladybird
Paropsis sp.

(a) Acacia leaves are the main food of this beetle but fortunately it rarely occurs in large enough numbers to do serious damage.

(b) The larvae also live on and chew acacia leaves.

(c) The pupae are usually hidden in the soil near the base of the food plant.

Adult length 1.2 cm
Family Chrysomelidae

93 Spotted leaf-eating beetle
Phola octodecimguttata
Vitex seems to be the main food plant of this handsome
18-spotted leaf-eating beetle. Sometimes plant damage is
sufficiently serious as to warrant chemical treatment.
Adult length 7 mm
Family Chrysomelidae

94 Irridescent leaf-eating beetle
Edusella glabra
These small, black beetles have either an iridescent green
or purple sheen. They are very hungry feeders on the
young leaves and shoots of a wide range of native plants.
They are difficult to catch and quickly fly away if disturbed.
Sometimes chemical control may be necessary.
Adult length 4 mm
Family Chrysomelidae

95 Hibiscus flower beetle
Aethina concolor
The adults often occur in large numbers in hibiscus
flowers or sometimes in magnolia flowers. They feed on
pollen and damage petals. Their larvae breed in fallen
flowers, which should be removed as a control measure.
Adult length 3–4 mm
Family Nitidulidae

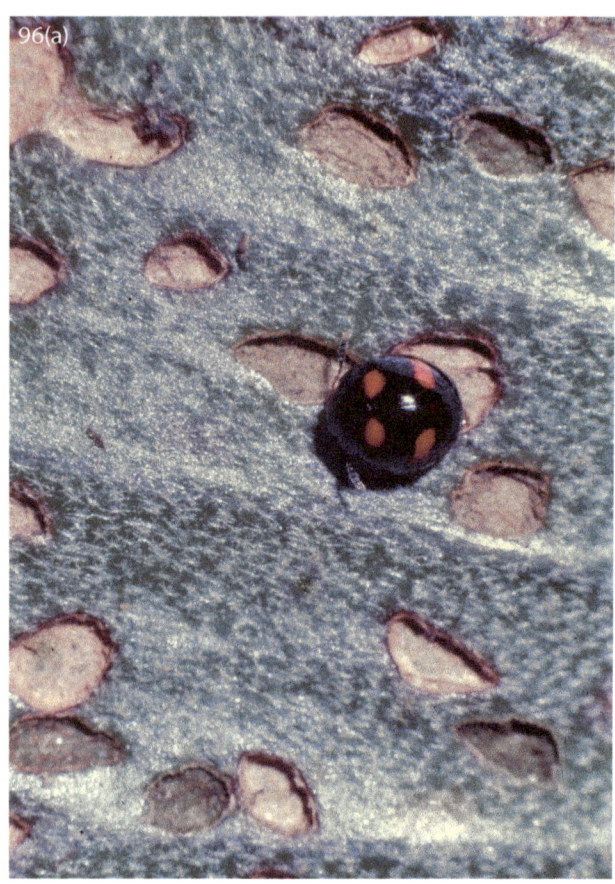

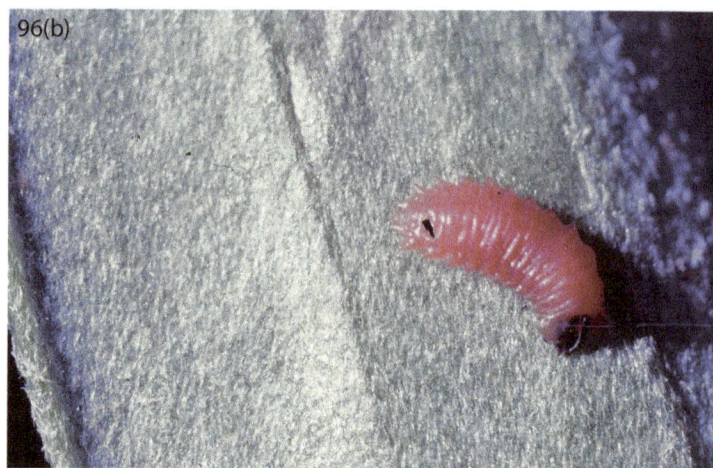

96 Staghorn fern window-spot beetle
Halticorcus platycerii
(a) The characteristic small window spots chewed by these tiny beetles are better known than the beetles themselves or their larvae. The spots do not extend right through the fronds, but leave a layer of clear or brown cuticle like a window.
(b) The tiny pink larvae burrow or mine within the fronds. Often, periodic collection and destruction by hand is sufficient to prevent serious damage but sometimes chemical treatment is required.
Adult length 3 mm
Family Chrysomelidae

97 Leaf-eating weevil
This damage is done by the feeding of an unidentified adult weevil beetle that is rarely seen because it feeds at night.
Adult length 5 mm
Family Curculionidae

98 Hibiscus flea beetle or red-headed flea beetle
Nisotra sp.
These flea beetles can occur in large numbers on hibiscuses and riddle the leaves with holes. They are called flea because when disturbed they jump like a flea. The best method of control is with a bucket containing a few centimetres of water and detergent, held under infested leaves which are then ruffled, causing the beetles to jump.
Adult length 6 mm
Family Chrysomelidae

99 Acacia flea beetle
Phyllotreta sp.
This is another species of flea beetle, but is smaller, black and harder to crush than other flea beetles.
Adult length 5 mm
Family Chrysomelidae

100 Christmas beetle
Anoplognathus spp.
(a) Several similar-looking species are called Christmas beetles because they are most active in November, December and January. In some years they congregate in large numbers on eucalypts and they may defoliate large areas of forest.
(b) They eat the leaves of eucalypts, melaleucas, leptospermums and, when food is short, other plants such as guava.
(c) The eggs are laid on or in the soil and the larvae (curl grubs or white grubs) live in the soil and, depending on the species, feed on dead vegetable matter or the live roots of plants for one or two years before pupating. (See also Illustration 243.) Chemicals are not very effective against adult beetles. Shaking the beetles to the ground and crushing is of slight benefit.
Adult length 2.5 cm
Family Scarabaeidae

101 Brown lousy beetle *Lepidota* sp.
This is one of several very similar-looking brown beetles that are active about the same time as the Christmas beetles. The adults are generally less damaging to plants than Christmas beetles, but the larvae are known as pasture white grubs.
Adult length 2.5 cm
Family Scarabaeidae

GRASSHOPPERS
See Chapters 2 and 3, also Illustrations 198, 199

102(a)

102(b)

102(c)

102(d)

102 Common garden grasshopper
Valanga irregularis
(a) Although there may be several different sizes and colours of short-horned grasshoppers in the garden, by autumn most are adults of this species.
(b) The eggs are laid in a mass or pod deep in the soil, and the individual eggs look rather like the juice sacs of an orange.
(c), (d) The nymphal stages are different colours, sometimes giving the impression that they are different species of insects. All stages are destructive leaf-eaters and, although periodic catching and destroying by hand helps to minimise plant damage, chemical treatment often has to be applied.
Adult length 9 cm
Family Acrididae

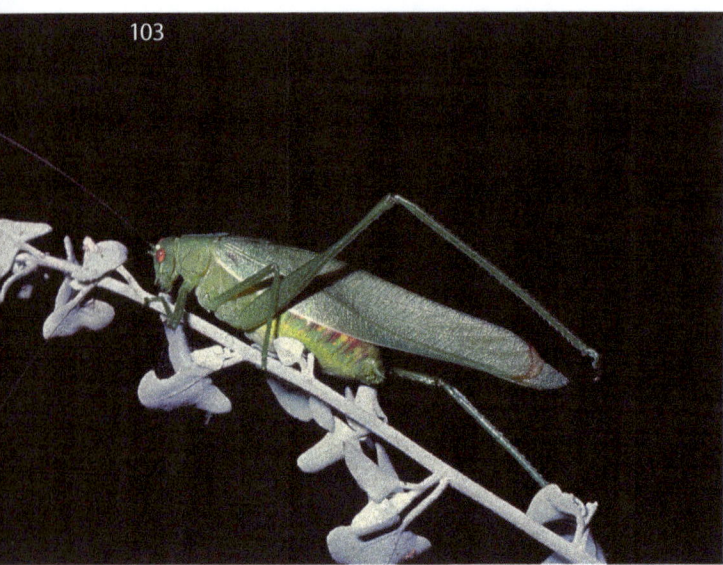

103 Long-horned grasshopper
Caedicia sp.
Wingless nymphs as well as winged adults of this common, destructive pest hide amongst leaves and feed on leaves, flowers, buds and fruit. Removal and destruction by hand is usually sufficient to keep damage to a minimum.
Adult length 4 cm
Family Tettigoniidae

104 Crested grasshopper
Alectoria superba
These very ornate long-horned grasshoppers prefer the open, sunny garden situation and usually a drier climate. Although they chew a few leaves and flowers, they are rarely present in sufficient numbers to be serious pests. If necessary they may be removed and destroyed by hand.
Adult length 7.5 cm
Family Tettigoniidae

PHASMIDS (STICK INSECTS)
See Chapters 2 and 3, also Illustrations 157 to 160

105 Phasmid
Anchiale austrotessulata
This fairly common species sometimes occurs in large numbers, particularly in gardens adjacent to bushland. All sizes of nymphs, as well as the adults, may be present at the same time and, unless they are thinned out, some plants may be partly defoliated. As can be seen with this mating pair, the fully winged male is much smaller than the short-winged female.
Adult length 22 cm
Family Phasmatidae

106 Phasmid
Tropidoderus sp.
This large and beautiful species may reach a length of 25 cm. Although often solitary, it is one of the plague species that sometimes occur in large numbers and defoliate forests.
Adult length 25 cm
Family Phasmatidae.

107 Phasmid eggs
Phasmid females drop their eggs at random and the hard, seed-like eggs may lie amongst the leaf litter for up to two years before hatching.
Egg size 3–4 mm
Family Phasmatidae

NEMATODES
See Chapters 2 and 3, also Illustration 247 and compare 332, 333, 336

108 Leaf nematode
Aphelenchoides sp.
The angular brown discolouration of parts of the fern
fronds is caused by microscopic nematode worms. They
move about on wet leaves and can be splashed from plant
to plant. The nematodes enter the leaves and feed
internally, causing the discolouration. The marks are
angular because the nematodes inside the leaf cannot
cross the larger veins. Leaf nematodes are difficult to
control even with specific chemicals.
Adult up to 1 mm
Family Aphelenchoididae

SEED EATERS
See Chapters 2 and 3

109 Seed-eating caterpillar

Sometimes ripe fruit, such as the *Hovea* fruit shown, is collected only to find that the contents have been eaten by a small caterpillar. This unidentified species is the larva of a moth. Control is difficult because the damage is done within the fruit. Seed should be cleaned soon after harvest and protected with Derris Dust.

Adult size 12–18 mm

Order Lepidoptera

110 Seed wasp

While these were normal-looking mature seeds and fruit at harvest, in the days following collection, minute adult wasps emerged. These included a seed-eating species, *Eurytoma* sp. (family Eurytomidae), and a parasitic species, *Eupelmus* sp. (family Eupelmidae). Some wasps cut a neat emergence hole through the fruit while some emerge from the seed after it has fallen from the fruit. Control is difficult because the damage is done within the fruit. Seed should be cleaned soon after harvest and protected with Derris Dust.

Adult length 1.5 mm

Order Hymenoptera

111 Tobacco beetle

Lasioderma serricorne

Seeds in storage, such as these palm seeds, may be completely eaten out by the larvae and adults of this small introduced beetle. It is commonly called the tobacco beetle because it is a serious pest of tobacco leaf and many other products in storage. Seeds in storage should be protected with Derris Dust.

Adult length 2 mm

Family Anobiida

112 Palm seed weevil

The adult weevil lays eggs in the ripe fruit while still on the palm and while in storage the larva hatch and eat the contents of the seeds.

Larva length 8–11 mm

Family Curculionidae

LEAF MINERS
See Chapters 2 and 3, also Illustrations 96, 108, 226

113 Cissus leaf miner
The scribble-like marks in the leaf are the feeding tunnels of a tiny larva of an unidentified insect that lives within the leaf. The diameter of the mine is at first quite small but it is enlarged as the larva grows and pupates, before the emerging adult breaks free.

114 Macadamia leaf miner
Acrocercops chionosema
In addition to macadamias, the larvae of the tiny moth that make these marks also attack new shoots on related plants such as buckinghamia and hakea. At first the mines are narrow and scribble-like but soon a larger blister-like area or blotch is mined. Sometimes it may be necessary to apply chemical treatment to a severe infestation.
Adult size 4–6 mm
Family Gracillariidae

115 Callistemon leaf miner
Heliozela sp.

(a) Leaf discolouration and shedding of some leaves are the main signs of this minute pest. The condition has been mistaken for fertiliser burn. If the red discolouration about the base or midrib of leaves is examined closely, the mines can be seen.

(b) The fully fed larvae cut oval-shaped pieces of leaf in which they pupate. These may be attached to the plant or nearby objects, or may fall to the ground.

(c) The adult is a minute moth. Sometimes the pupae are parasitised by a tiny green wasp. Severe infestations may warrant use of chemicals.

Adult length 2 mm
Family Heliozelidae

FLIES
See Chapters 2 and 3, also Illustrations 1, 6, 139 to 142, 286, 287, 328

116

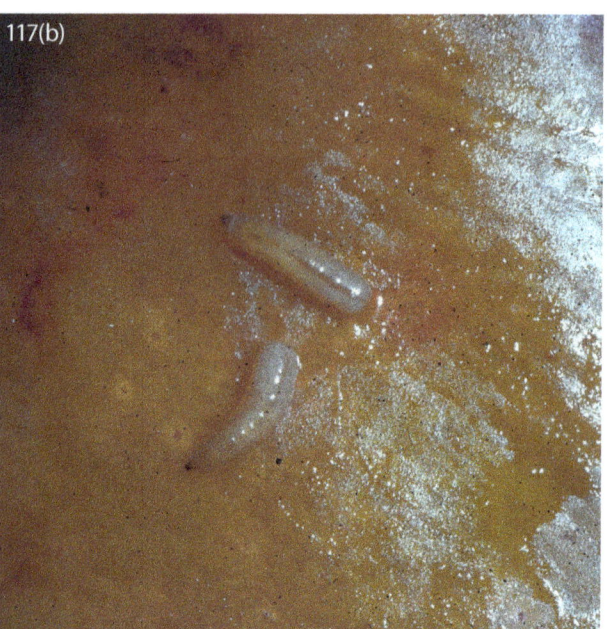

117(b)

116 Hibiscus flower flies
These unidentified flies congregate in large numbers in open hibiscus flowers damaging the petals. As hibiscus flowers last only a day or two, it is practical to use a household pest pressure pack spray on the open flowers. One or two days' treatment will usually eliminate the pest.
Adult length 3 mm
Order Diptera

117(a)

117(c)

117 Queensland fruitfly
Bactrocera tryoni
(a) The so-called Queensland fruitfly is native to the east coast as far south as Victoria, particularly the higher rainfall areas. However, hot, wet summers favour its regular occurrence whereas the dry summers of the south and the drier climate of inland districts limit its activities and it may become obvious only during favourable seasons. The wasp-like adult flies are often attracted to blossoms and they breed in many native and exotic fruits.
(b) The maggots tunnel within fruit causing fungal infection and a soft rot.
(c) The introduced weed *Solanum mauritianum* (wild tobacco tree), besides being classed as a serious pest because it can invade established pasture, is also a major host for fruitflies.
Adult length 7 mm
Family Tephritidae

118 Island fruitfly
Dirioxa pornia
This fruitfly is of minor importance compared to the
so-called Queensland fruitfly because it can penetrate
only split or damaged fruit.
Adult fly 5.5 to 8.5 mm
Family Tephrididae

119 Tabanid flies
This is a distinct group of blood-sucking flies known as
March flies or horse flies, which harass humans as well as
causing a great deal of hysteria among livestock.
Adult length 1–4 cm depending on species
Family Tabanidae

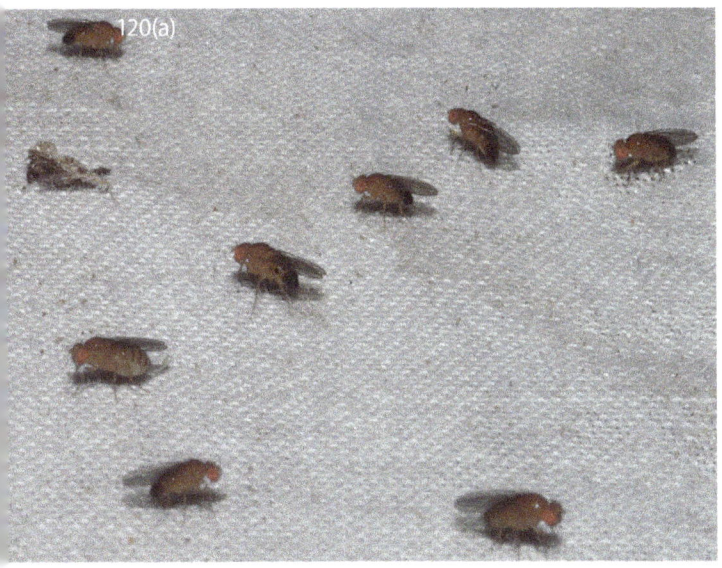

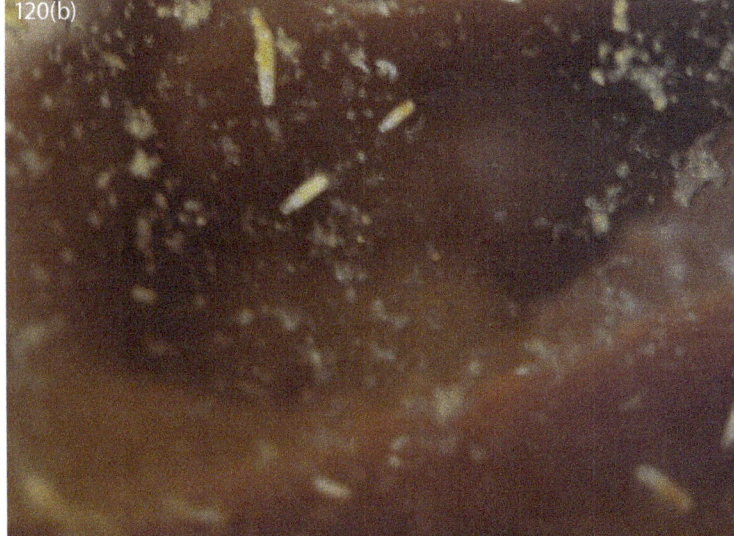

120 Vinegar fly
Drosophila sp.
(a) Often wrongly called fruitflies, the vinegar fly is
associated with over-ripe fruit and in fact feeds on the
yeasts associated with decaying fruit.
(b) The very tiny maggots are hard to see and are usually
overlooked by those enjoying over-ripe fruit.
Adult length 3 mm
Family Drosophilidae

121 Wallaby fly
Ornithomya sp.
This blood-sucking ectoparasite of mammals and birds is sometimes found on gardeners – having been left behind in the garden by a native animal. They are flattened, have wings but rarely fly and are superbly adapted to making rapid movement through fur, feathers or clothes. A fly can appear out of a sleeve and a few seconds later around the neck or the other sleeve.
Adult length 9 mm
Family Hippoboscidae

LESS HARMFUL, HARMLESS AND BENEFICIAL SMALL ANIMALS

Fortunately, many of the small animals that are often seen on flowers, fruit, leaves and shoots are not pests. Some do so little damage that their removal is not warranted. They may have other virtues such as beautiful or unusual larvae or adults. Others are quite harmless and many are, to a greater or lesser extent, beneficial in that they feed on pest species.

For further information relating to the following illustrations, the reader should refer to the appropriate section in Chapters 2 and 3.

MOTHS AND BUTTERFLIES
See Chapters 2 and 3, also Illustrations 65 to 81, 109, 114, 115, 221 to 226, 245, 283

122 Banksia hawk moth, double-headed hawk moth
Coequosa triangularis
(a) The large and beautiful moth is Australia's largest
hawk moth.
(b) This unusual and beautiful large caterpillar is relatively
rare and should not be regarded as a pest. It rarely can be
found other than singly and the few leaves it eats would
not be missed from a healthy banksia or persoonia plant.
Wingspan 15 cm
Family Sphingidae

123 Orchard butterfly
Papilio aegeus aegeus
(a) Even though the caterpillars of this beautiful butterfly eat the leaves of citrus, including the native species, it is a pleasure and a privilege to have such an exquisite animal floating around the garden.
(b) Provided the plant on which the caterpillar is feeding is healthy and well foliaged, the few leaves eaten by a small number of these caterpillars does little harm. If too many caterpillars are present or the plant is too small, the excess caterpillars could simply be removed by hand and squashed.
(c) The unusual pupae can sometimes be seen on or near food plants.
Wingspan 10 cm
Family Papilionidae

124(a)

124(b)

124(c)

124 Tailed emperor butterfly
Charaxes sempronius
(a) This beautiful butterfly breeds on leguminous plants, mostly acacias. It sips nectar from flowers or can occasionally be seen sucking from decaying fruit.
(b) The caterpillars, too, are very attractive and, because they usually occur in only small numbers in a garden, they rarely warrant removal. The few leaves they eat are a small price to pay for the beautiful butterflies.
(c) The pupae may be attached to the food plant or to nearby plants or objects.
Wingspan 8 cm
Family Nymphalidae

125 Emperor moth
Opodiphthera sp.
(a) The size, brown colour and prominent wing eye spots are quite distinctive.
(b) The large caterpillars of this beautiful moth are studded with stinging bristles. These larva feed on a variety of different trees.
Moth wingspan 9–10 cm
Family Saturniidae

126 Bizarre looper or zigzag caterpillar
Eucyclodes pieroides
(a) This unusual little caterpillar is often mistaken for a fragment of dead leaf. It does very little damage, does not occur in large numbers and feeds on a wide range of plants.
(b) The beautiful moths are attracted to light.
Wingspan 3 cm
Family Geometridae

127(a)

127(b)

127(c)

127 Pretty moths and butterflies
Chelepteryx chalepteryx
(a) Caterpillar length to 7 cm, moth wingspan to 10 cm
Family Anthelidae
(b) Brown moth
Neola sp.
Occasionally a beautiful moth or butterfly will be seen in the garden or may fly into the home at night. You may never know what its caterpillar looks like or where it feeds. Surely this is sufficient reason why all caterpillars should not be indiscriminately killed on sight.
Wingspan 10.5 cm
Family Notodontidae
(c) Jezebel
Delias sp.
The caterpillars of this colourful butterfly feed on mistletoes.
Wingspan 6.5 cm
Family Pieridae

128 Ant-tended caterpillar – imperial blue
Jalmenus evagoras
(a) The ants are quite tolerant of the adult butterflies as they tend the caterpillars and pupa.
(b) The ants guard the caterpillar and the pupae in return for their secretions.
Adult wingspan 4–4.5 cm
Family Lycaenidae

129 Cup moth caterpillar
This beautiful cup moth caterpillar feeds on macadamias and carries sharp, stinging, non-retractable spines. Its adult and habits are similar to the following species.
Family Limacodidae

130(a)

130(b)

130(c)

130 A cup moth
Doratifera sp.
(a) This small and distinct group of moths have unusual and often beautifully coloured, slug-like caterpillars.
(b) The cup-like pupae of some are attached to twigs; others are hidden in the bark or at the base of the host plant in the soil.
(c) The larvae or caterpillars often carry stinging hairs which, in some species, can be retracted or, if the larva is disturbed, erected. Occasionally they occur in sufficient numbers as to warrant removal or at least thinning out. If necessary, these caterpillars could be carefully knocked off rather than sprayed with chemicals.
Wing span 3 cm
Family Limacodidae

131 Mottled cup moth
Doratifera vulnerans
(a) This caterpillar, shown with its spines erected, often feeds on eucalypts and melaleucas and it stings very sharply if touched. It is sometimes called the 'Chinese junk' because of its square shape.
(b) The adult has the typical squat shape and downy appearance of other cup moths.
Wingspan 3 cm
Family Limacodidae

132 Saunders case moth
Metura elongatus
(a) Saunders case moths live a solitary existence and feed on a wide range of plants. They rarely warrant removal because they occur in small numbers. Case moths caterpillars live within the protection of portable, silken cases or bags usually disguised with pieces of leaf or twig. Pupation takes place within the bag. With some species both the males and females are winged and emerge from the bag; in others the female is wingless and remains within the bag.
(b) This newly hatched caterpillar has constructed a silken bag in which to live and shelter. This is enlarged as the caterpillar grows.
Adult case length 10–12 cm
Family Psychidae

133 Silky case moth (ribbed)
Hyalarcta huebneri
This species of case moth can sometimes be seen on eucalypts but rarely warrants removal. Unlike many of the other species of case moth, it does not decorate its bag with fragments of leaf or twig.
Adult case length 4–5 cm
Family Psychidae

134 Stick-bound case moth
This harmless case moth covers its portable home with twigs. It can sometimes be seen in gardens.
Case length 2–4.5 cm
Family Psychidae

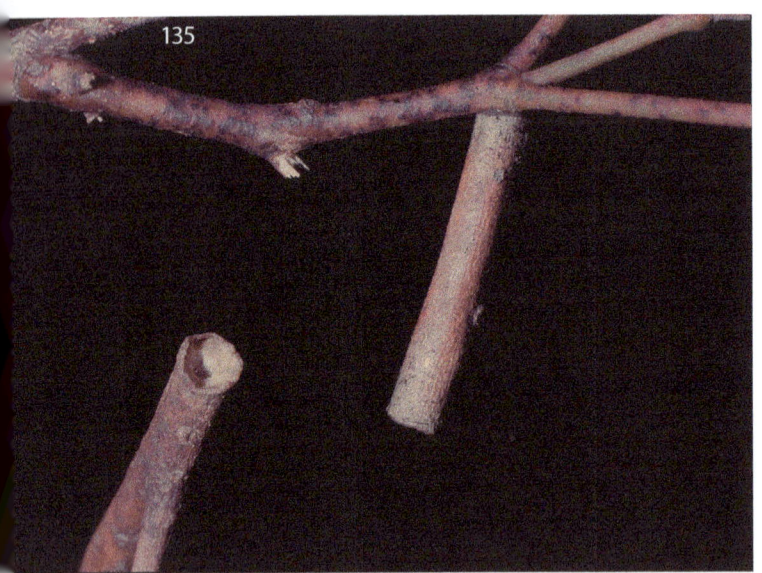

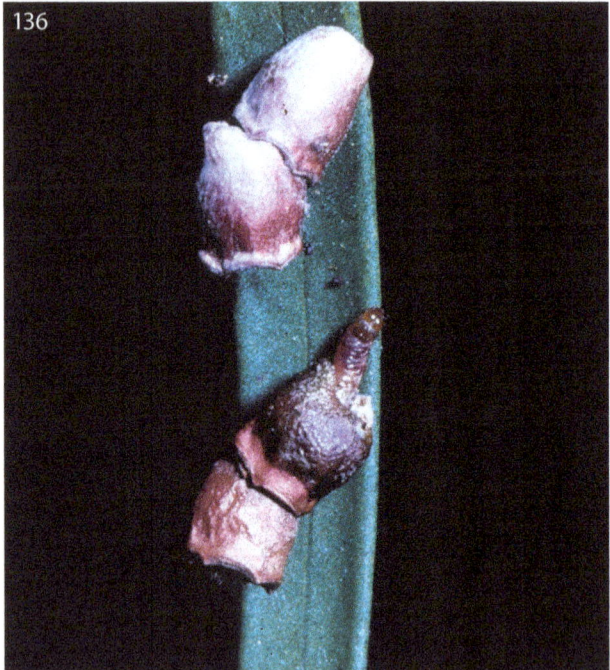

135 Twig-hollowing moth
Hemibela sp.
This unusual moth larva chews off a small piece of twig of suitable diameter and hollows it out to make a portable home. When it grows too large for the home it chews off another section of a larger twig. It is not a pest, but its unusual habits are interesting.
Twig shelter length 20–28 mm
Family Oecophoridae

136 Capsule case moth
Spilonota constrictana
This tiny caterpillar eats the contents of immature callistemon capsules and joins two capsules together to make a portable case for itself. It turns into a small moth. It can scarcely be regarded as a pest.
Adult length 4 mm
Family Tortricidae

ANTS

See Chapters 2 and 3, also Illustrations 128, 137, 236, 237

137 Ants and scale insects
Ants feed eagerly on the sugary excretions of several sap-sucking insects, such as the scale shown. They will, in fact, carry insects such as aphids, mealy bugs and scales from plant to plant and sometimes build shelters around them or care for them in their nests.
Ant length 7 mm
Family Formicidae

138 Rattler ants
Polyrhachis sp.
These ants are not aggressive and have no sting, but if provoked may bite with their jaws. They bind leaves together or use hollow stems for homes and, if disturbed, bang the surface of their home with their rear end, creating a rattling sound, presumably to frighten off predators.
Adult length 6 mm
Family Formicidae

FLIES
See Chapters 2 and 3, also Illustrations 1, 6, 116 to 121, 286, 287, 328

139 Tachinid fly
Rutilia sp.

(a) These and several other related bristly flies are often seen feeding on the nectar in flowers. Tachinid flies play a very important part in controlling the large caterpillars and other large insects that are too large for wasps to tackle. The adult fly lays its eggs either directly on the host or in places frequented by the host.

(b) Instead of a moth hatching from the moth pupa, several fly pupae usually drop out before hatching into tachinid flies.

(c) The white spots on the body of this large caterpillar are the eggs of a tachinid fly. The larvae or maggots tunnel and feed within the host.

Adult length 1.4 cm
Family Tachinidae

140 Common hoverfly
Simosyrphus grandicornis
(a) These very beneficial flies are so called because of their habit of hovering motionless in the air for several seconds at a time. Their larvae are important in controlling aphids, thrips, scale insects and other small pests.
(b) The tiny, white, oval eggs are deposited amongst the pest insects (there is an egg close in front of the larva shown) and the green larva or maggot is feeding hungrily on aphids.
Adult length 9 mm
Family Syrphidae

141 Hoverfly
Melanostoma sp.
This is another species of hoverfly, shown at rest on the flower of a native iris.
Adult length 1 cm
Family Syrphidae

142 Spider fly
Sarcophaga reposita
The tent spider *Cyrtophora moluccensis* constructs a tent-like web with the eggs suspended vertically down the centre of the tent. Note the parasitic fly at the top of the egg mass laying its maggots on the spider eggs.
Adult fly length 8–12 mm
Family Sarcophagidae

WASPS
See Chapters 2 and 3, also Illustrations 3 to 5, 110, 278 to 282

143 Parasitised aphids
Note the emergence holes in the empty aphid skins. The aphids have been parasitised by a tiny wasp whose larva have eaten the contents of each aphid, pupated within the aphid and emerged, making the escape hole.
Aphid length 2–2.5 mm, wasp adult length 5–7 mm

144 Scale wasp parasite
Several minute wasps parasitise scale insects and aphids. This tiny iridescent unidentified wasp is one of several species that play a very important part in control of scales and aphids.
Adult length 1 mm

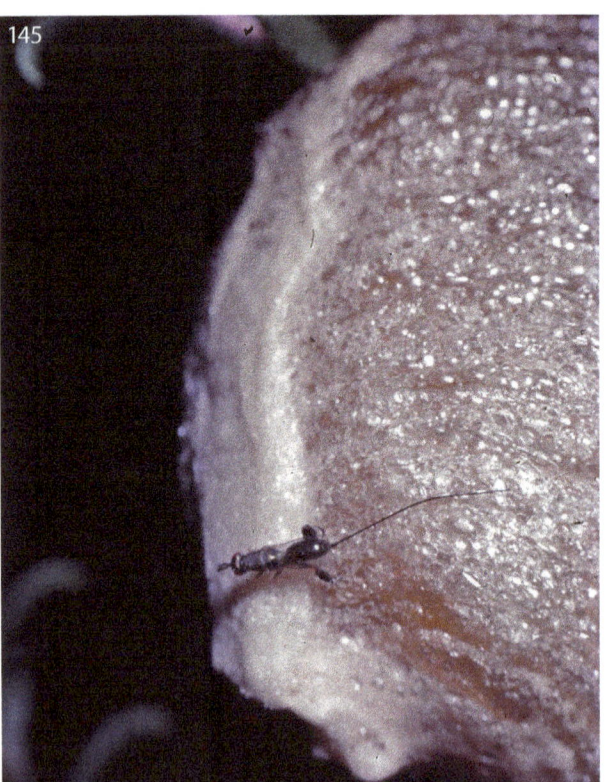

145 Mantis egg parasitic wasp
Podagrion sp.
Mantis eggs are frequently heavily parasitised by this minute wasp. The very long, thread-like ovipositor enables the wasp to penetrate the frothy protective covering around the egg mass and deposit its eggs into the mantis eggs.
Adult length (with ovipositor) 4 mm
Family Torymidae

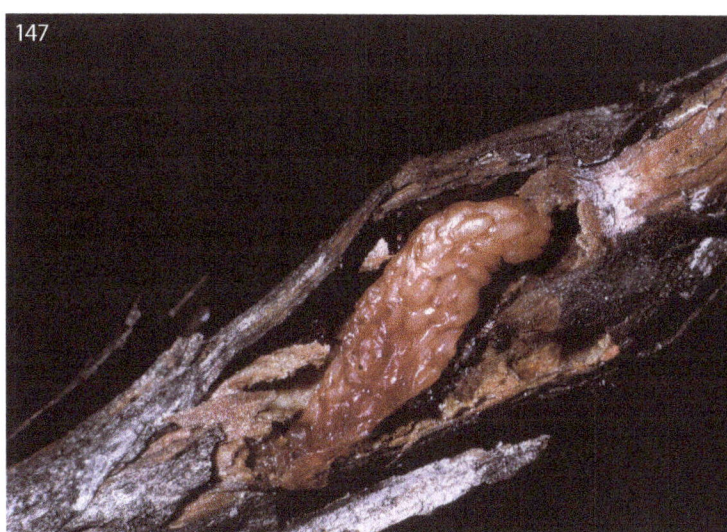

146 Pollinating wasp
The pollination of most Australian native plants is carried out by the many species of flies and wasps. This unidentified wasp is one of a few species in which the wingless female is attached to the male and transported in copula to flowers. They are important pollinators.
Adult length males 2 cm, females 1.5 cm
Family Tiphiidae

147 Parasitised larva
The exposed larva, probably of a beetle borer, has been parasitised by a wasp, its body eaten out by the wasp maggots which have now formed a mass of pupas about to hatch.
Parasitised larva 2.8 cm, wasp pupa 2 mm

148 Parasitised caterpillar and emerged pupae
Apanteles sp.
(a) The caterpillar has been parasitised by a tiny wasp. The wasp laid a special kind of egg, which developed into a large number of minute maggots. The contents of the caterpillar are gradually eaten away and it dies slowly as the fully fed maggots leave the caterpillar and form a mass of white pupae nearby.
(b) Minute black wasps have emerged from the pupae surrounding the dead caterpillar.
Adult length 1.5 mm
Family Braconidae

149 Lawn caterpillar wasp
Lissopimpla sp.
So often when gardeners see this wasp active on their lawns, they rush out with the spray because they believe the wasp is eating the grass. In fact, the wasp is highly beneficial because it is seeking out and inserting its eggs into the grass caterpillars and pupae.
Adult length (with ovipositor) 3–4 cm
Family Ichneumonidae

150 Parasitised pupa
Brachymeria sp.
The hole in the side of the pupa has been made by the emergence of several small wasp parasites. Their larvae have eaten the contents of the butterfly pupa.
Adult length 2 mm
Family Chalcididae

151 Bark borer larva and parasite
Gotra sp.
(a) The maggot has hatched from an egg laid on or near a bark-boring caterpillar upon which it has been feeding externally.
(b) The adult of the maggot is this ichneumon wasp.
Adult length 7 mm
Family Ichneumonidae

PAPER WASPS
See Chapters 2 and 3

153 Paper wasp
Polistes sp.
These flattened papery nests are usually built on branches or under the eaves of buildings. The wasps sting savagely if closely approached and are best removed from accessible places. Nevertheless they are beneficial garden insects.
Adult length 29–23 mm
Family Vespidae

152 String paper wasp
Ropalidia gregaria
All paper wasps can sting if disturbed, some quite sharply. However, before you remove them, it is interesting to know that they feed their young on masticated caterpillar and therefore they play a very important part in reducing the number of caterpillars in the garden.
Adult length 1 cm
Family Vespidae

MANTIDS
See Chapters 2 and 3

154(a)

154(b)

154 Common mantis
Archimantis latistyla
(a) Praying mantids are among the most ferocious creatures on Earth today. Tease a mantis and you will soon find how ferocious it can be. A 15 cm specimen can defy a cat or a dog once it is aroused. They are slightly beneficial, feeding on a wide range of insects, some of which may be pests.
(b) The frothy or spongy egg mass shown with newly emerged young is usually attached to a twig or leaf.
Adult size 10–12 cm
Family Mantidae

155 Blue-spot mantis
Orthodera ministralis
There are many species of mantids. They are called praying
mantids because, while waiting in ambush for a victim,
they hold their raptorial forelegs in a way that suggests
they are praying. This small species displays a bright blue
spot on the inside of each foreleg when it is extended.
Adult length 5 cm
Family Mantidae

156 Bark mantis
This unidentified mantis nymph (wings not yet developed)
is well camouflaged on the bark of a eucalypt tree.
Nymph length 6 cm
Family Mantidae

PHASMIDS
See Chapters 2 and 3, also Illustrations 105 to 107

157 Acrophylla stick insect
Acrophylla titan
(a) When disturbed these large stick insects fluff their colourful wings to frighten off predators.
(b) These very large stick insects or phasmids sometimes attain a length of 36 cm. They eat the leaves of a wide range of plants, but fortunately they are usually solitary insects that rarely warrant removal. The few leaves eaten by one or two of these insects in a garden will not be missed.
Adult length 30+ cm
Family Phasmatidae

158 Extatosoma leaf insect

Extatosoma tiaratum

(a) This unusual phasmid may reach a length of 16 cm. It is a beautiful and relatively rare species that does not warrant eradication. Although sometimes called a leaf insect, the true leaf insects belong in another section of Phasmatidae and are rare in Australia. They are recorded from the Atherton Tableland in North Queensland.

(b) The male is smaller and slender.

(c) They sometimes develop camouflage to mimic their surroundings. This nymph was living on a lichen-covered shrub.

(d) The eggs are seed-like and may lie on the soil for a couple of years before hatching. Rain stimulates hatching.

Adult length 14–16 cm

Family Phasmatidae

160 Terrestrial phasmid
This unidentified species is sometimes found among sticks
on the ground.
Adult length 10–14 cm
Family Phasmatidae

159 Goliath stick insect
Eurycnema goliath
This large spectacular insect is sometimes kept as a pet. If
disturbed it fluffs its colourful wings to frighten predators.
Adult female length 20 cm
Family Phasmatidae

BEETLES

See Chapters 2 and 3, also Illustrations 8, 86 to 101, 111, 112, 216 to 220, 241 to 244, 252, 253

161 Ladybird scymnodes
Apolinus lividigaster
(a) This brown, spiky animal shown here feeding on aphids is the larva of a ladybird beetle.
(b) The adult is this small black and yellow ladybird beetle. Ladybird is a corruption of 'Our Lady's bird' and was applied first to a species of coccinellid that appeared quite 'miraculously' to control an outbreak of pests in European vineyards in the days long before pesticides. The adult also eats aphids.
Adult length 3 mm
Family Coccinellidae

162 Ladybird harmonia
Harmonia conformis
(a) This pupa of a ladybird beetle and similar pupae of other species of ladybirds can often be seen attached to the leaves of garden plants.
(b) The adult is this handsome spotted ladybird. Both the larvae and the adult beetles are predators that feed on pests. There are many species of predatory ladybirds, varying in colour from black and iridescent green to yellow or orange with black stripes or spots.
Adult length 6 mm
Family Coccinellidae

163(a)

163(b)

163 Ladybird coccinella, transverse ladybird
Coccinella transversalis
(a) This common predatory species lays its yellow eggs in clusters near or amongst colonies of aphids, psyllids or other pests.
(b) The black and pale yellow larvae eat large numbers of aphids and other pests, but are frequently mistaken for and killed as plant-eaters.
Adult length 5 mm
Family Coccinellidae

164 Ladybird cryptolaemus
Cryptolaemus montrouzieri
(a) This adult, an extremely important beneficial ladybird, feeds on scale, aphids and mealy bugs.
(b) The larvae of cryptolaemus look like mealy bugs and generally are exterminated as such. The two larvae shown here are feeding on the cottony egg sacs of pulvinaria scale. In contrast to the sluggish habits of mealy bugs, cryptolaemus larvae move relatively quickly looking for pests [See Illustrations 29 and 208].
Adult length 4 mm
Family Coccinellidae

165 Ladybird stethorus, mite-eating ladybird
Stethorus sp.
This tiny black ladybird beetle is about half the size of a pin head (1 mm in length) and is an important but inconspicuous predator of mites. It can be seen here eating a red spider mite. Note also the young pale-coloured scale insects.
Adult length 1 mm
Family Coccinellidae

166 Mantis egg beetle
Thaumaglossa nigricans
(a) The adults are these small, dark, mottled beetles that eagerly eat mantis eggs.
(b) The yellowish, hairy larvae feed hungrily on mantis eggs. They almost totally consume the entire egg mass and frothy covering.
Adult length 2.5 mm
Family Dermestidae

SPIDERS
See Chapters 2 and 3, also Illustrations 230, 259 to 262

167 Blossom spider
Thomisus spectabilis
This crab-like spider resembles in colour the flowers in which it hides. It seizes insects that visit flowers, and therefore its food includes beneficial and harmless insects, as well as pests.
Size 14 mm
Family Thomisidae

168 Jumping spider
Oxyopes sp.
This very agile spider hides amongst leaves and seizes passing insects. It can spring quite large distances. As with all spiders, it is beneficial because it eats many pest insects.
Adult length 15–20 mm
Family Oxyopidae

169 St Andrew's cross spider
Argiope aetherea
This spider usually remains motionless, attached to the centre of its web during the day, but at night it traps flying insects including many egg-laden adult pests. This one has caught a grasshopper.
Adult leg span 3–5 cm
Family Araneidae

170 Bird-dropping spider
Celaenia excavata
This species, shown here with its eggs, spends the day disguised as a bird-dropping. At night it is busily engaged in trapping insects.
Adult size 14 mm
Family Araneidae

171 Long-tailed spider
Arachnura sp.
This unusual spider is easily passed over as a fragment of twig or leaf during the day. At night it traps other spiders.
Adult length 15–20 mm
Family Araneidae

172 Jewel spiders
Poecilopachys australasia
(a) Known also as the two-spined spider, this is one of several decorative and harmless species called jewel spiders because they resemble jewellery.
Adult size 10–12 mm
Family Araneidae
(b) Jewel spider
Austracantha minax
This is another species with a wide distribution.
Adult size 10–12 mm
Family Araneidae

PREDATORY BUGS
See Chapters 2 and 3 also plant bugs and Illustrations 228, 288

173 Assassin bug
Pristhesancus papuensis
(a) This predatory bug captures and sucks the juice from anything that comes along, even fingers if you are careless. It often hides in flowers and captures beneficial insects, as well as pests. It can be seen with its proboscis inserted into a fly.
(b) Eggs are laid in a cluster and glued to a leaf.
Family Reduviidae

174 Predatory bug
Amyotea hamata
Predatory bugs can capture and immobilise insects much larger than themselves, as shown by this nymph feeding from a grasshopper. The toxin in their saliva acts very quickly on their victims.
Adult length 1.4 cm
Family Pentatomidae

176 Brown predatory bug
This common, unidentified predatory bug moves actively among plants and preys on various insects including pests.
Adult length 18 mm

175 Predatory bug
In spite of its small size (about 1.5 mm) this predatory bug captures and sucks from small caterpillars, psyllids and other pests. It is an important, but inconspicuous, natural control. A nymph only is shown. Note the white waxy secretions of the grevillea bud psyllid.
Adult length 1.5 mm
Family Miridae

LEAF AND PLANTHOPPERS
See Chapters 2 and 3, also planthoppers Illustrations 23, 24, 203 to 205

177 Spittle bug

Philagra sp.

(a) The frothy mass under which the soft-bodied nymphs shelter looks just like a blob of spittle, hence the common name.

(b) When the froth is removed the nymph can be seen. The froth is produced by the nymphs as a protective cover from dryness and predators.

(c) At the conclusion of the final nymph stage, the adult emerges. It can be seen here not yet fully expanded, as it emerges from the froth.

(d) The long-nosed adult is seen here after hardening and colouring. Both the nymphs and adults suck sap, but are not regarded as serious pests. Hosing will remove many of them from the branches.

Adult length 8 mm

Family Cercopidae

178 Froghopper
Chaetophyes sp.
The nymphs of this planthopper construct small calcareous tubes attached to twigs. They live in and feed from within these tubes immersed in their own liquids and are thus protected against drying out and predators. They are not serious pests but, if removal is considered necessary, they may be rubbed or pruned off.
Adult length 5 mm
Family Clastopteridae

179 Giant mealy bug
Monophlebulus sp.
This very large, solitary mealy bug is occasionally to be found in gardens, but is very readily attacked by predators and parasites. It is more of a curiosity than a pest.
Adult length 2.5 cm
Family Monophlebidae

180 Melaleuca hairy gall
Sphaerococcus sp.
These unusual hairy galls are common on twigs of paper bark melaleucas such as *Melaleuca quinquenervia*. They are caused by a species of sphaerococcus that develops and feeds within the gall. They are more of a curiosity than a problem.
Gall diameter 1.5–2 cm
Family Pseudococcidae

LACEWINGS
See Chapters 2 and 3, also Illustration 251

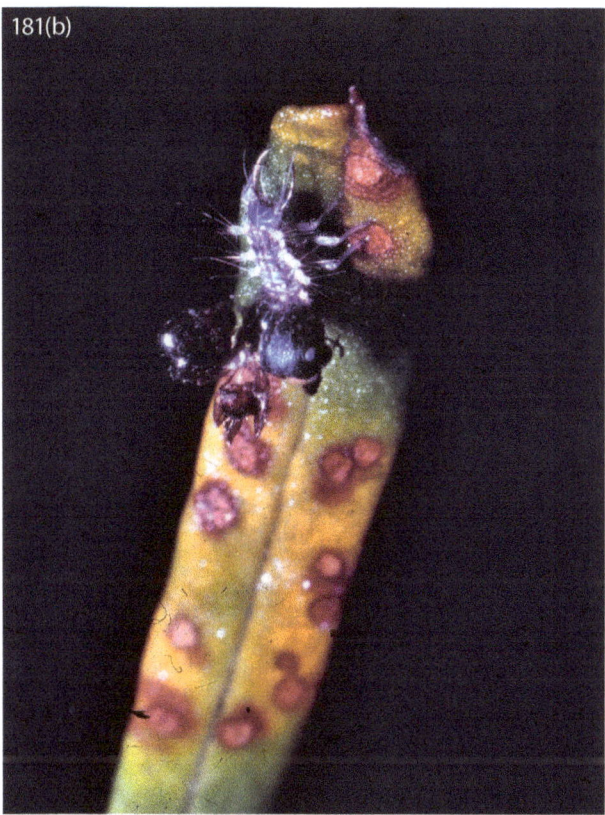

181 Green lacewing
Mallada signatus
(a) This very common species frequently comes indoors to lights. Both the adults and the larvae are highly beneficial predators of pests such as mites, aphids, scales, caterpillars and moths.
(b) The tiny larva lives on leaves and branches, covers its body with the empty skins of its victims and is not often noticed.
Wingspan 2.5 cm
Family Chrysopidae

182 Yellow lacewing
Nymphes myrmeleonoides
This large yellow species is occasionally seen sheltering amongst plants. As with other species, both adults and larvae are valuable predators.
Wingspan 7.5 cm
Family Nymphidae

183 Spotted lacewing
This is the adult of another often seen unidentified species of lacewing.
Wingspan 7 cm
Family Osmylidae

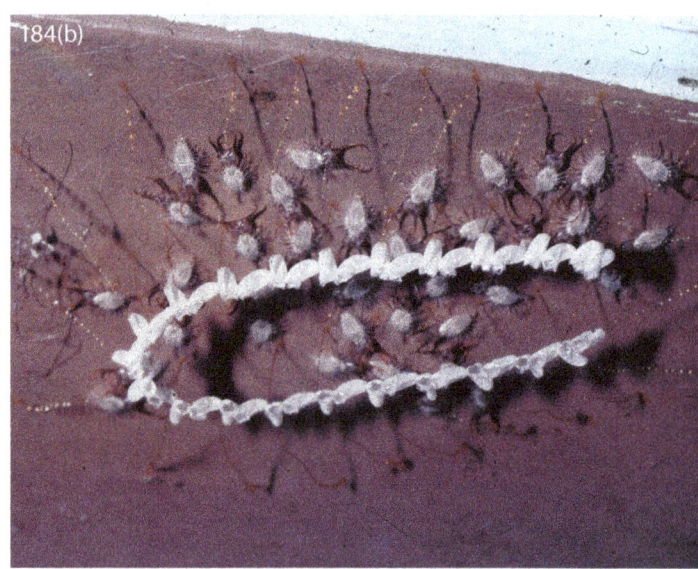

184 Lacewing eggs
(a) So ferocious are the tiny larvae that hatch out of lacewing eggs that if they were laid close together the first to hatch would eat its fellows. However, the female lays each egg on a long stalk to aid dispersal.
(b) Some species of lacewing lay their eggs in a characteristic U-shaped pattern.
Order Neuroptera

185 Lacewing larvae

(a), (b), (c) The best known of the lacewing larvae are the ones that make dust traps in dry soil and are known as ant lions. (See Illustration 251.) Less known are the inconspicuous species, the larvae of which live on plants. Some of these disguise their bodies with the empty bodies of their victims and pieces of bark and other debris. The next time you notice a minute piece of litter moving, have a close look; it will probably be a lacewing larva. Shown are three of the species that can be found on plants.
Larva length (a) 4–6 mm (b) 4–6 mm (c) 5–8 mm
Order Neuroptera

186 Lacewing larva and pupa

(a) This lacewing larva shown here with a captured moth lives on the trunks of trees and disguises its body with pieces of bark.

(b) Its pupa is spherical in shape and attached to bark where it is difficult to detect.

Order Neuroptera

BEES
See Chapters 2 and 3

188 Teddy bear bees
Amegilla bombiformis
Although most of the bees that can be seen on flowers are the introduced honey bees, several native species also live in gardens, including this large, solitary species that nests in the soil. At night they 'roost' on twigs in sheltered spots as shown here, often holding on with their mouthparts while all legs clean the body.
Adult length 2 cm
Family Apidae

187 Blue-banded bees
Amegilla sp.
Although these are solitary bees, they often appear grouped together because they may make their individual nests in close proximity.
Adult length 1.5 mm
Superfamily Apidae

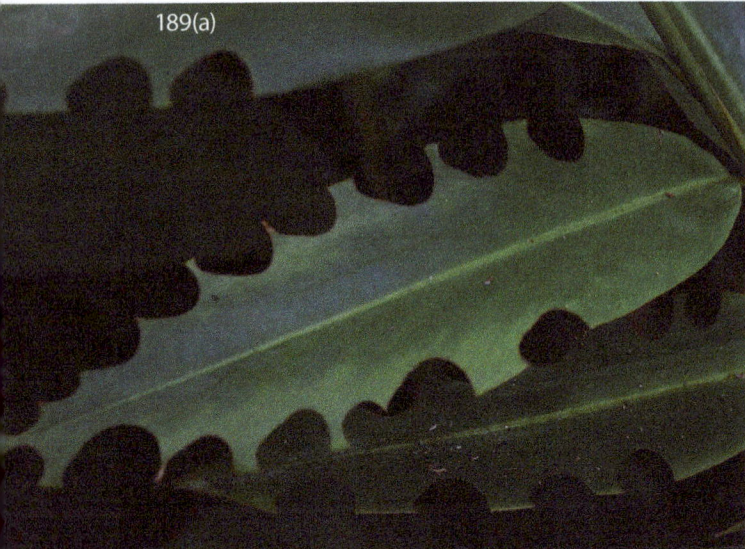

189(a)

189(b)

189(c)

189 Leaf cut by leaf-cutter bee
Megachile sp.

(a) These are best known by the neat semicircles they cut out of the leaves of plants such as rose and ginger or related plants. The leaf pieces are used to line tunnels or burrows in rotten logs or soil and to construct cells in which a ball of pollen and nectar and a single egg are placed.

(b) Tubes of leaf pieces in a rotting log.

(c) The adult female bee.

Adult length 10–14 mm

Family Megachilidae

190 Bees in flowers
Lasioglossum lanarium
The males of this native bee frequently congregate overnight in flowers.
Adult length 10–12 mm
Family Halictidae

191 Firetailed resin bees
Megachile sp.
(a) This bee with a red abdomen crawls into sheltered places – in this case, a box of seed packets – and fills spaces with resin or wax in which it constructs cells containing pollen and nectar; it then deposits its eggs in the cells.
(b) The adult bee.
Adult length 2–2.5 cm
Family Megachilidae

192 Sugarbag bee
Tetragonula hockingsi
Adult length 5 mm
Family Apidae

CICADAS
See Chapters 2 and 3, also Illustrations 215, 257, 329

193 Cicada
(a) This unidentified cicada nymph is leaving the soil after two or three years, to hatch into an adult.
(b) The nymph attaches to a branch or bark of a tree, splits down the back and slowly emerges as an adult cicada as its wings expand.
(c) An emerged and hardened adult is shown beside the empty shell from which it emerged.
Adult length 4 cm
Family Cicadidae

194 Bladder cicada
Cystosoma saundersii
These unusual, large leaf-like insects shelter amongst the leaves of trees and shrubs and give forth a deep-throated 'song' just about dusk. Although they suck sap, they are harmless insects.
Adult length 5 cm
Family Cicadidae

MITES
See Chapters 2 and 3, also Illustrations 11, 53 to 59, 212, 235, 263

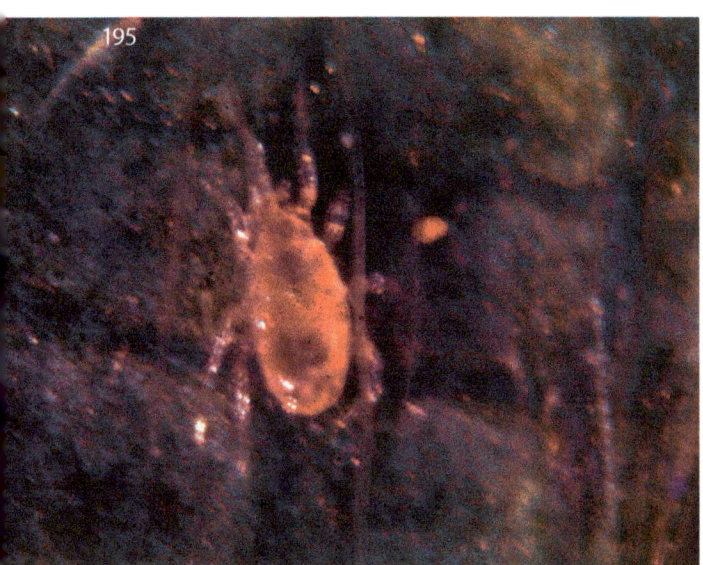

195 Predatory mite
Phytoseiulus persimilis
Although microscopic in size, this predatory mite feeds on other mites and small insects and is a very important natural control. The eight legs of the mite can be seen.
Microscopic size
Family Phytoseiidae

196 Domatia and mites
The leaves of some plants have minute hollows on the undersides at the junction of veins. These are called domatia. Their purpose is somewhat obscure, but frequently several types of predatory mites can be seen living in these 'homes'. The illustrated domatia are pale edged and the reddish mite is near domatia just above the midrib.

CRICKETS
See Chapters 2 and 3, also Illustrations 255, 256

197

197 Tree cricket
Hadrogryllacris sp.
Tree crickets feed on a mixed diet of leaves, flowers and mantis and other insect eggs, and sometimes on other insects or the debris from insects. The nymphs of this species, as shown here, bind leaves together to form a shelter. The adult frequently lives under loose bark and in similar shelter, closer to the ground than the nymphs.
Adult length 5.5 cm
Family Gryllacrididae

GRASSHOPPERS
See Chapters 2 and 3, also Illustrations 102 to 104

199 Hooded brown katydid
This unusual, unidentified long-horn grasshopper is
possibly part predatory.
Adult length 5 cm
Family Tettigoniidae

198 Predatory grasshopper
This small, long-horn grasshopper likes damp, shady
places and can often be found hiding among leafy plants.
Its spiny front legs indicate it is a predator, catching and
eating other insects as well as eating some plant tissue.
Adult length 3–4 cm
Family Tettigoniidae

COCKROACHES
See Chapters 2 and 3, also Illustration 254

200 Arboreal cockroach
Ellipsidion sp.
(a) This small cockroach hides among the leaves of trees and shrubs. Adult fruit flies and flea beetles are recorded as being among the food it eats, but possibly only as a scavenger of dead or dying insects, and therefore it must be regarded as a beneficial insect.
(b) It is quite a colourful insect.
Adult size 1.5 cm
Family Ectobiidae

SNAILS AND SLUGS
See also Chapters 2 and 3, also Illustrations 248 to 250, 272

201 Tree snail
Helicarion sp.
Snails inhabit sheltered positions on or near the soil. This native species can occasionally be found in gardens, probably brought in from the bush in ferns. It lives and feeds amongst the leaves of trees or shrubs metres above the ground and mainly eats decaying plant litter. It covers its shell with its mantle while it is moving about.
Shell size 1–1.5 cm
Family Helicarionidae

202 Rainforest snail, giant panda snail
Hedleyella falconeri
This large, dark snail is native of rainforests where it feeds on rotting plant debris and fungi. It is often encountered in gardens adjacent to forest and it does not harm plants.
Shell size up to 9 cm
Family Caryodidae

PESTS ASSOCIATED WITH TWIGS AND SMALL BRANCHES

The insects illustrated in this section more or less confine their activities to the leafy twigs or the small stems just below the leaves. Some of these insects have evolved particularly interesting habits and among these should be included some of the sap-sucking species, some gall-forming species and the ringbarking branch borers. While some species of weevil damage twigs in a minor way by feeding on bark, serious damage by the larvae or grubs to the roots of infested plants may pass unnoticed, even after plants have been killed.

For further information relating to the following illustrations, the reader should refer to the appropriate section in Chapters 2, 3 and 5, and Appendix I.

PLANTHOPPERS
See Chapters 2 and 3, also Illustrations 23, 24, 177, 178

203 Planthopper
Some planthoppers live together in colonies, which may contain several stages of nymphs and the adults. All stages suck sap and, if the insects are present in large numbers, plants may become stunted.
Adult length 10–12 mm
Superfamily Fulgoroidea

204 Planthopper
Two adults and three last-stage nymphs can be seen, as well as their ant attendants.
Adult length 6 mm
Family Cicadellidae

205 Introduced Planthopper
Aconophora compressa
This insect was introduced from America in 1995 to assist with the control of the serious weed lantana. It has been found to attack a range of other introduced plants such as fiddlewood, duranta, jacaranda, clerodendron and native plants such as eremophila and myoporum. It causes branch dieback and stunting.
Adult length 7 mm
Family Membracidae

SCALE INSECTS
See Chapters 2 and 3, also Illustrations 29 to 37, 137, 164

206 Lac scale
Austrotachardia sp.
The hard, brown scale shown here among the darker coloured capsules of a melaleuca secretes a hard protective covering of shellac. This scale seems to prefer a dry climate and it infests a wide range of native shrubs. Infested shrubs die back severely. In Asia some species of this scale are farmed for production of shellac.
Scale length 4 mm
Family Kerriidae

207 Cottony cushion scale
Icerya purchasi
This large, soft-bodied scale is capable of multiplying at a very fast rate. It can attack a wide range of plants. Fortunately, it is usually kept under control by other native insects such as ladybird beetles. Occasional outbreaks occur when pesticides kill off its predators. It excretes copious amounts of honeydew. In 1887, this scale had been inadvertently introduced to California from Australia and almost wiped out the Californian citrus industry. With the introduction of a ladybird beetle (*Rodolia cardinalis*) in 1889, the pest was brought under control in 15 months and the foundation of biological control was firmly laid.
Scale length (with egg sac) 1.5 cm
Family Monophlebidae

208 Nigra scale
Parasaissetia nigra
This introduced scale attacks a range of plants, and its sugary excretions are a frequent cause of sooty mould. Note the countless young scales and the *Cryptolaemus* larvae feeding on young scales.
Scale size 4 mm
Family Coccidae

209 Latania scale
Hemiberlesia lataniae
Dieback of branches and even death of entire plants may
be the first sign of infestation by this inconspicuous scale.
A close inspection and scraping of the bark on sick plants
may be necessary to determine its presence. The severe
effect of this scale is caused by its toxic saliva. Grevilleas
are particularly susceptible.
Scale size 2 mm
Family Diaspididae

210 Tick or pearl scale
Cryptes baccatus
This repulsive-looking, large scale attacks mainly acacias. It
sometimes forms large clusters like fully engorged ticks on
the twigs and small branches. Heavy infestations cause
dieback.
Scale size 4–5 mm
Family Coccidae

211 Rice bubble scale
Eriococcus coriaceus
Eucalypts are the main host for this scale, particularly
species that are unsuitable for the district where they are
being grown. It often causes severe dieback and death of
heavily infested plants. It excretes copious amounts of
honeydew and a thick growth of sooty mould is associated
with it, as are ants.
Scale size 3–4 mm
Family Eriococcidae

MITES

See Chapters 2 and 3, also Illustrations 11, 53 to 59, 195, 196, 235, 263

212 Witch's broom mite
This deformed growth is fairly common on species of casuarina. Its cause has not been conclusively determined, but microscopic mites can consistently be recovered from young bunches of growths. Deformed shoots may be pruned off and burnt.

GALLS
See Chapter 2, also Illustrations 1 to 11, 180, 247, 302, 318, 332, 333, 336

213 Casuarina gall
Cylindrococcus spiniferus
(a) There are many very interesting gall-forming insects, including this species that galls casuarina. The galls closely resemble the fruit and can sometimes be mistaken for seeds. There are no fruit in the illustration, only galls.
(b) The red, shapeless animal in the opened gall is an adult female. Note the tiny red nymphs, which crawl away to form new galls nearby. Early pruning of young galls may help to reduce infestations.
Adult length 8 mm
Family Eriococcidae

214 Eucalyptus apiomorphid gall
Apiomorpha pileata
These unusual galls are frequently covered with the mealy, white, waxy secretions of the causal insect, which is closely related to mealy bugs. They rarely assume pest proportions.
Gall size 6–8 cm diameter
Family Eriococcidae

CICADAS
See chapters 2 and 3, also Illustrations 193, 194, 257, 329

215 Cicada damage

Aleeta curvicosta

(a) Eggs are laid in a zigzag cut and on hatching the nymphs fall to the ground where they burrow deeply in the soil and suck from roots for several years.

(b) Sometimes wrongly called locusts, these insects are well known for the shrill, ear-piercing noise made by the adults. The female here is making a zigzag cut on the branch with her ovipositor.

Adult length 4–5 cm

Family Cicadidae

BEETLES
See Chapters 2 and 3, also Illustrations 8, 86 to 101, 111, 112, 161 to 166, 241 to 244, 252, 253

216(a)

216(b)

216 Weevil damage
(a), (b) This type of damage is fairly common on melaleuca, callistemon and related shrubs, and it is caused by the feeding of an adult weevil. It could be an indicator that the roots of the plant are under attack by the larvae of the weevil. The insect causing the damage is rarely seen (see Illustrations 241, 242).
Family Curculionidae

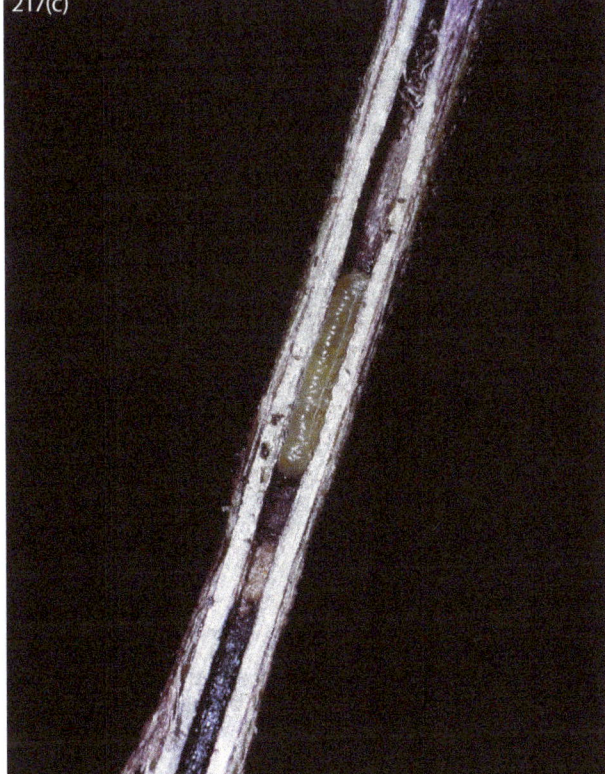

217 Eucalyptus seedling borer
Rhadinosomus lacordairei
(a) This unusual, slender weevil attacks seedlings of
eucalyptus such as *Eucalyptus shirleyi* and *E. miniata*.
(b) A small hole chewed at a stem node may be the only
early sign of attack.
(c) The larvae tunnel within the small stems causes
dieback and misshapen plants.
Adult length 8 mm
Family Curculionidae

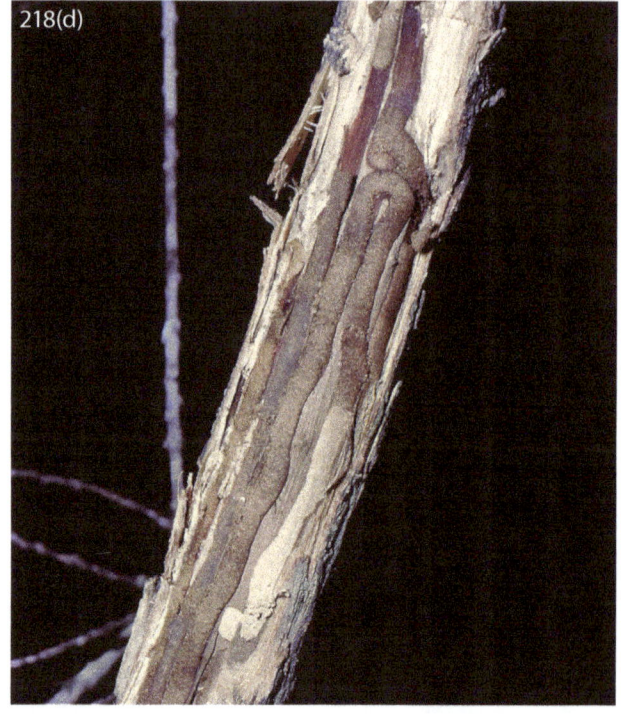

218 Callistemon dieback borer
Rhytiphora armatula
(a) Although the symptom of dieback occurs high on affected plants (mostly callistemon species), there may be little evidence of the cause on or near the dead twigs.
(b) The infested stem may be slightly swollen and tend to sucker (many small shoots).
(c) Sometimes the yellowish larvae or grubs can also be found boring under the bark.
(d) Removal of bark well down the main stem may reveal the feeding tracks in the sapwood.
(e) The adult is this handsome longicorn beetle.
Adult length 2.5 cm
Family Cerambycidae

219(a)

219(b)

219(c)

220

219 Ringbark longicorn beetle
Rhytiphora piperitia
(a) This beetle does a very neat job of ringbarking.
(b) Eggs are laid in small semicircular cuts above the ringbark. By ringbarking branches, the beetle is following a principle used by horticulturists whereby a branch is cinctured to promote starch enrichment of the wood for easier propagation of difficult plants. The beetle has found that starch-enriched wood is good food for its larvae.
(c) The developing grub a few weeks after egg laying.
Adult length 2.4 cm
Family Cerambycidae

220 Longicorn beetle
Ancita marginicollis
The long antenna typical of longicorn beetles is clearly illustrated by this showy species.
Adult length 1.8 cm
Family Cerambycidae

MOTH BORERS
See Chapters 2 and 3, also Illustrations 77, 224, 225, 245

221 Moth borer
The dark-coloured caterpillar hides by day in a shallow hole chewed into the branch at a fork. As an added protection, it covers the area around the fork with webbing disguised with its own droppings. At night, it feeds on nearby leaves, sometimes bringing pieces of leaf back to the shelter, where they are fastened to the outside. It attacks a range of plants including grevillea, hakea and melaleuca.
Adult length 3–4 cm
Order Lepidoptera

222(a)

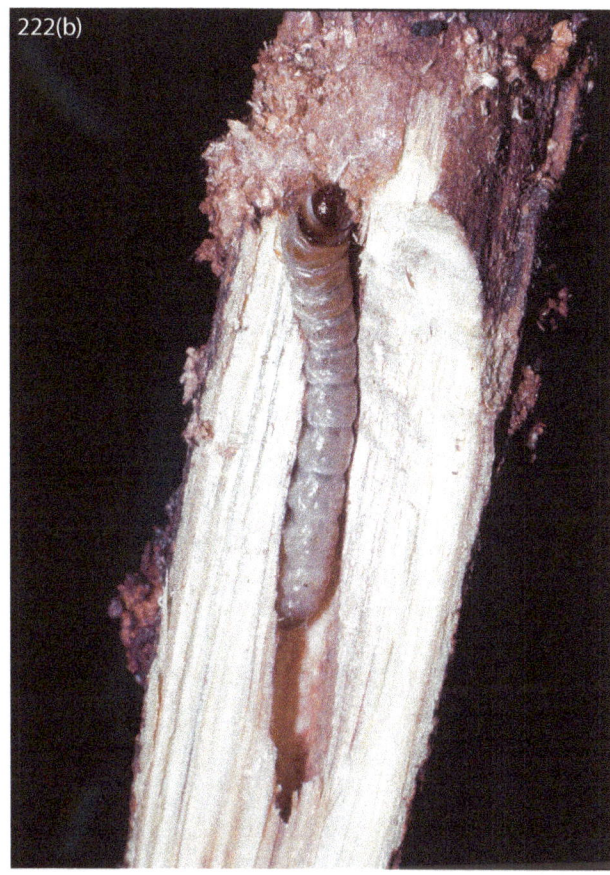

222(b)

222(c)

222 Moth borer

Cryptophasa sp.

(a) Acacias are the principal food plant of this moth borer. The caterpillars attack the main branches and trunks, feeding mainly on the bark and cambium layer. The area of their activity is covered with webbing, camouflaged with their own droppings and chewed fragments of wood. Often the trunk on which they feed is ringbarked and the upper portion dies.

(b) The caterpillar also chews a deeper shelter hole into the wood where it eventually pupates.

(c) Finally, the silver coloured moth emerges.

Adult length 3 cm

Family Oecophoridae

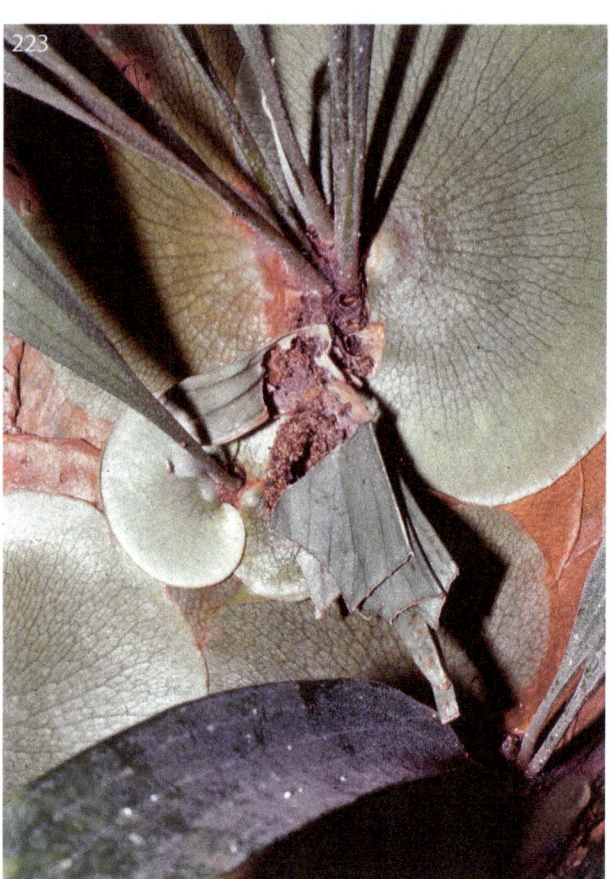

223 Fern moth borer
A caterpillar, the larva of a moth, tunnels and shelters within the peaty bases of elkhorns, staghorns and hare's-foot ferns. At night it feeds on nearby fronds and it often carries portions of fronds back to its tunnel, where they can be seen protruding from the shelter hole or fastened outside.
Adult moth length 4–5 cm
Family Oecophoridae

PESTS AND BENEFICIAL ANIMALS ASSOCIATED WITH TRUNKS AND MAIN BRANCHES

The main branches and trunks of trees and shrubs are not attacked by as many pests as the upper parts. However, a variety of small, harmless or beneficial animals find shelter and food there, particularly on or under the bark of eucalypts.

For further information relating to the following illustrations, the reader should refer to the appropriate section in Chapters 2 and 3.

MOTHS AND BUTTERFLIES
See Chapters 2 and 3, also Illustrations 65 to 81, 122 to 136, 221 to 223, 245, 283

224(a)

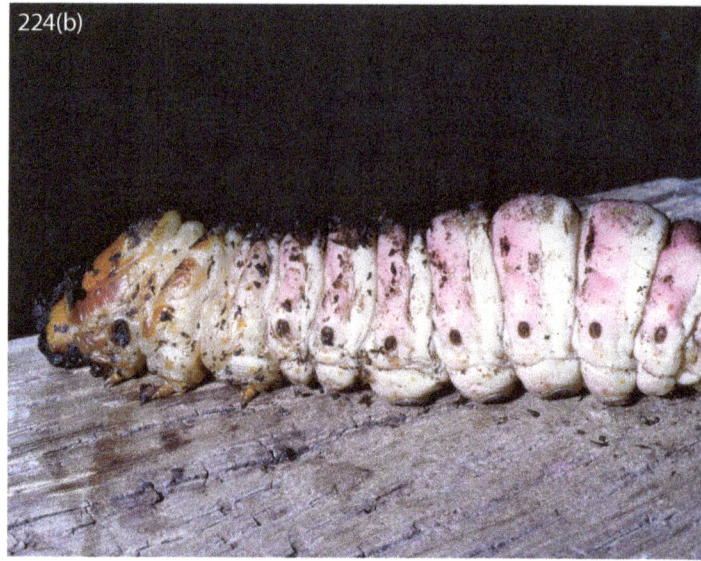

224(b)

224(d)

224(c)

224 (see also opposite) Xyleutes, giant wood moth
Endoxyla cinereus
(a) Usually the first sign of the presence of this large insect is an augur-like emergence hole, which at times has provoked accusations of tree poisoning.
(b) The large, wood-boring caterpillar feeds up and down in the centre of the trunk with little external sign of its presence and it may take two years or more to develop and pupate.
(c) Sometimes the empty pupal case remains protruding from the emergence hole.
(d) The moth sometimes rests beside the emergence hole. Note the smaller breather or frass hole.
(e) The large moth.
Wingspan 20 cm
Family Cossidae

225 Eucalyptus bark borer, webbing borer
(a) This moth borer is very common at times on the trunk of smooth-barked eucalypts. They are most plentiful in newly developed subdivisions where eucalypts are the only part of the original bush that is retained. Possibly, this abundance is because of the destruction of the natural ecosystem with its parasites and predators. The caterpillars construct web- and droppings-covered 'runways' on the surface of the bark and at night they chew the surface of surrounding bark.
(b) The dark caterpillars each have a shallow shelter hole chewed in the bark, where they eventually pupate.
Adult moth 2–2.5 cm
Family Oecophoridae

226 Scribbly gum moth
Ogmograptis scribula
(a) The scribbly gum gets its name from the characteristic scribble-like marks or mines made on its bark by the mining larvae of a small moth. The moth larvae in no way harm the tree and cannot be regarded as pests. In fact, the scribbles add a great deal of charm to the already beautiful gum tree.
(b) A close-up of the scribbles.
Family Bucculatricidae

BUGS
See Chapters 2 and 3, also Illustrations 41 to 52, 173 to 176, 288

227 Shovel nose bug
Stenocotis depressa
This harmless bug is a master of disguise both in its extremely flat shape and in its colouration. It blends perfectly with the bark on which it feeds.
Adult length 1.8 cm
Family Cicadellidae

228 Bark bugs
Poecilometis sp.
These are nymphs of predatory bugs that live on the trunks of eucalypts. In the nymphal stages they usually hide in groups under loose pieces of bark.
Adult length 2 cm
Family Pentatomidae

EARWIGS
See Chapters 2 and 3

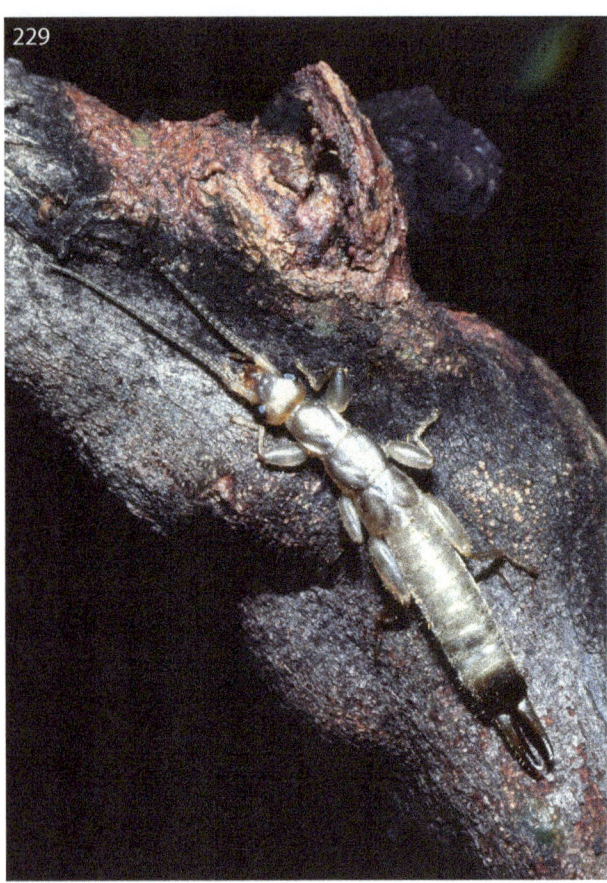

229

229 Earwig
Earwigs can frequently be found sheltering during the day under loose pieces of bark on the trunks of trees. The rear end of the insect is equipped with a very distinctive pair of forceps. The earwig illustrated is a nymph (wings not developed).
Adult length 2 cm
Order Dermaptera

SPIDERS
See Chapters 2 and 3, also Illustrations 167 to 172, 259 to 262

230 Bark spider
Several kinds of spider are commonly found sheltering
under loose bark. Some are very flattened to adapt to their
peculiar habitat. They are valuable predators of insects
and should not be disturbed unnecessarily.
Adult length 4.5 cm
Family Sparassidae

CLASS ARACHNIDA
See Chapters 2 and 3

231 Scorpion
Although scorpions are generally regarded with feelings of dread, the recorded observation of the remains of cockroach just eaten is proof of their beneficial feeding habits. They hide under loose bark or logs or stones.
Adult length (with claws) 3 cm
Order Scorpiones

232 Pseudoscorpion
These harmless or slightly beneficial tiny animals may include some plant pests in their diet. Superficially, they resemble true scorpions or at least the front ends of each are similar, but they lack the long stinging 'tail' of true scorpions. Pseudoscorpions can often be found under loose bark.
Adult length (with claws) 6 mm
Order Pseudoscorpiones

REPTILES
See also Illustrations 273 to 277

233 Gecko
Strophurus intermedius
This beautiful, nocturnal lizard hides under loose pieces of bark or in rock piles during the daylight. At night it actively hunts insects and it is highly beneficial.
Length 15 cm
Family Diplodactylidae

234 Gecko
Oedura sp.
Another nocturnal lizard that hides under loose bark during the day but hunts at night.
Length 12–16 cm
Family Gekkonidae

235 Spiny-tail gecko
Diplodactylus sp.
This was photographed in a garden where the barks of rose bushes and citrus trees were heavily infested with brevipalpid mites. Note the discoloured and splitting bark. This fascinating gecko remained motionless on the branches during the day but actively ate mites at night.
Length 7–10 cm
Family Diplodactylidae

PESTS AFFECTING COLLARS AND ROOTS

Several insects attack plants at ground level or below ground and some pests, such as soil nematodes, attack only the roots. Some below-ground pests have a severe detrimental effect on plants but, because the pest is out of sight, correct diagnosis can be difficult. When the cause is in doubt, it is advisable to remove dead plants from the ground carefully for a detailed examination of the collar and roots. Removal for examination should not be delayed too long or the damaged bark or roots may decay, obscuring the signs of damage.

For further information relating to the following illustrations, the reader should refer to the appropriate section in Chapters 2, 3 and 5, and Appendix I.

ANTS
See Chapters 2 and 3, also Illustrations 128, 137, 138

236 Coastal brown ant
Pheidole megacephala
Ants are among the most persistent and insidious of garden pests. They carry sap-sucking pests such as aphids, scales and mealy bugs from plant to plant, sometimes thereby spreading virus diseases. They also excavate around stems and roots, loosening plants in the soil and frequently they establish colonies of mealy bugs or coccids on the roots.
Worker length 1.5–4.5 mm
Family Formicidae

237 Fire ant
Solenopsis invicta
This is yet another of the increasing number of invaders in recent years. It is a sobering thought that no country has succeeded in eradicating this very serious pest. The adult ants in each nest are of various sizes.
Adult length 2–6 mm
Family Formicidae

COCCIDS AND MEALYBUGS
See Chapters 2 and 3, also Illustrations 38 to 40, 179

238 Root coccids

These insects, similar to mealy bugs, suck sap from the roots of plants, but are often overlooked because they are out of sight. If crushed, they look like congealed blood. They are usually ant-attended, and the activities of ants are often a good indication of their presence. Sometimes chemical control is necessary.

Adult length 2 mm

Family Pseudococcidae

239 Bulb mealy bug

Bulbous plants such as the native calostemma and exotic bulbs may be attacked by mealy bugs, which work their way down the leaf bases to the bulb in the soil. The leaves on affected plants come up mottled and distorted, and the symptom resembles that of a virus disease. Clusters of mealy bugs of various sizes can often be seen if the leaf bases are pulled apart.

Adult length 4 mm

Family Pseudococcidae

APHIDS
See Chapters 2 and 3, also Illustration 22

240 Root aphids (or pemphigids).
(a) The only signs of these serious pests were poor vigour of annual flower plants and a few ants present. Careful removal of a few plants made the problem obvious.
(b) These were woody shrub plants, both in pots and in the soil. They were decidedly slow-growing and had poor colour and, although an examination of roots revealed some small patches of 'cotton wool', it took close examination to eventually locate the culprits.
Size 2–3 mm
Family Pemphigidae

BEETLES

See Chapters 2 and 3, also Illustrations 8, 86 to 101, 111, 112, 161 to 166, 216 to 220, 252, 253

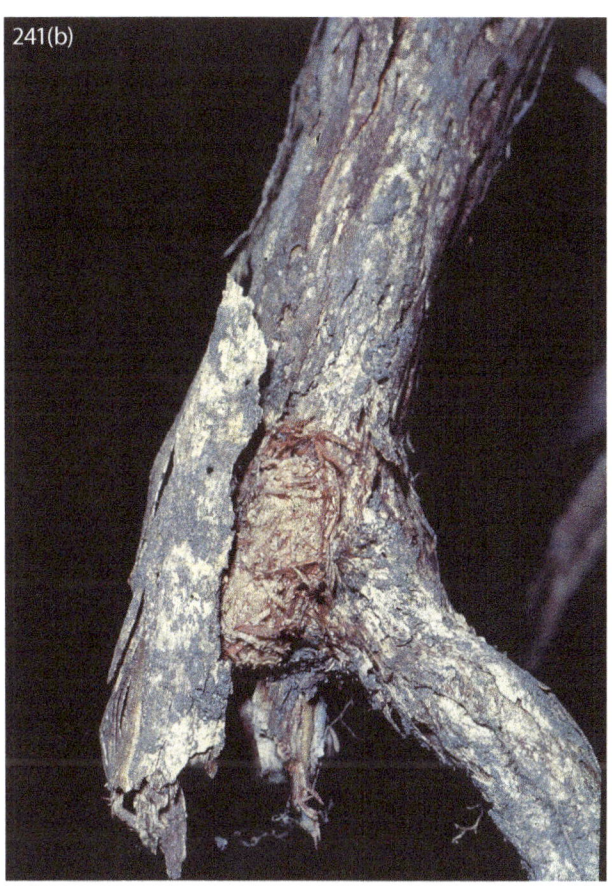

241 Root weevil
Aterpus griseatus
(a) Infested plants may die quite suddenly, often just as they begin to bloom. By carefully digging out a plant and hosing off the soil, you may find that the plant has been ringbarked below ground. Feeding damage to the twigs on top of a plant, as illustrated here with the adult weevil, is often an indication that eggs have been laid at the base of the plant.
(b) Sometimes the pupae can be seen in balls of chewed wood in depressions chewed in the roots.
Adult length 1.2 cm
Family Curculionidae

242 Backhousia root weevil
(a) Trees with yellow leaves in a plantation indicated a problem. Yellow trees were loose in the ground.
(b) Careful removal showed the extent of below-ground damage.
(c) Unidentified weevil larva from the base of the stem.
(d) Weevil pupae.
Family Curculionidae

243 Curl grub

(a) Several species of beetle have almost identical larvae; these include the Christmas beetles and the brown or lousy beetles [Illustrations 100, 101]. Some of these species damage stems and roots as illustrated here – others feed only on plant litter.

(b) A typical curl grub.

Grub length up to 6 cm

Family Scarabaeidae

244 Staghorn beetle

Xylotrupes sp.

The grubs that commonly inhabit compost heaps are larger than ordinary curl grubs, are covered with short hairs and have more prominent spots down the sides. Pupae form in fist-sized knobs of soil or compost. Late-stage orange pupa show the physical characteristics of the black adult. Adults often chew bark from branches of poinciana trees.

Adult beetle length 3–4 cm

Family Scarabaeidae

MOTHS AND BUTTERFLIES
See Chapters 2 and 3, also Illustrations 65 to 81, 109, 114, 115, 122 to 136, 221 to 226, 283

245(a)

245(b)

245(c)

245 White caterpillar
(a) This very destructive caterpillar lives in the soil and chews the bark off roots and stems.
(b) The pupa is also formed in the soil.
(c) Hatching of the pupa is stimulated by rainfall and the purplish moths can be seen in car headlights.
Moth wingspan 5–6 cm
Order Lepidoptera

TERMITES
See Chapters 2 and 3

246

246 Termites
Nasutitermes sp.
Although often called white ants, termites are not related
to true ants. Several species are found in gardens,
particularly where old stumps or roots from the original
trees still exist. Often these termites attack homes and
construction materials in contact with the soil; sometimes
they attack garden trees and shrubs. Some species
construct mud-covered tracks up the bark of eucalyptus
trees to dead branches where they feed. Termites also
perform the very important function of disposal of forest
litter and return of nutrients to the soil.
Size 5–6 mm
Family Termitidae

NEMATODES

See Chapters 2 and 3, also Illustrations 108 and for comparison 332, 333, 336

247(a)

247(b)

247 Root knot nematode
Meloidogyne sp.
(a) The root-galling nematodes are the best known of these microscopic, parasitic worms, simply because the galls are more obvious than the discoloured, stunted or rotted roots caused by other groups of the nematodes. Soft-wooded plants, such as daisies and mint bushes, generally form larger galls than hard-wood plants. Nematode galls may provide easy access for root-rotting fungi.
(b) A potted plant showing extensive galling of roots.
Size generally microscopic
Phylum Nematoda

SNAILS AND SLUGS
See Chapters 2 and 3, also Illustrations 201, 202, 272

248 Garden snail
Helix aspersa
The common garden snail was introduced to Australia and has become a serious pest of gardens and small-crop farms.
Shell diameter to 3.5 cm
Family Helicidae.

249 Orchid snail
Zonitoides arboreus
This tiny introduced species frequently lives in pots of orchid compost and feeds on the tips of roots and on young shoots.
Shell diameter 2 mm
Family Gastrodontidae

250 Garden slug
Deroceras sp.
This introduced species is a serious pest of gardens and small-crop farms.
Slug length 2–4 cm
Family Agriolimacidae

SMALL ANIMALS ASSOCIATED WITH SOIL, COMPOST AND SHELTER

Many small animals live very sheltered and secret lives, buried in the soil or hidden within compost or under pots, stones or other objects. Compost heaps in particular are teeming with all manner of small animals, a few of which feed on plants and many of which play a significant part in decomposition of plant material into humus. Others are important predators of pests such as snails and plant-feeding insects.

For further information relating to the following illustrations, the reader should refer to the appropriate section in Chapters 2 and 3.

NEUROPTERA
See Chapters 2 and 3, also Illustrations 181 to 186

251(a)

251(b)

251(c)

251 Ant lion
(a) These pit traps are constructed in loose dry soil by the larvae of a species of lacewing in order to trap ants and any other small animals that happen to fall in.
(b) The larva or ant lion hides in the dust near the bottom of the trap and quickly seizes any victim, dragging it under the soil where it is sucked dry.
(c) The adult is often mistaken for a dragonfly or damselfly but has an unmistakable weak fluttery flight.
Wingspan 6.5 cm
Family Myrmeleontidae

BEETLES
See Chapters 2 and 3, also Illustrations 8, 86 to 101, 111, 112, 161 to 166, 216 to 220, 241 to 244

252 False wire worm
These peculiar, hard, wire-like animals are the larvae of beetles that live in compost and decaying vegetation. False wire worms can be minor pests of roots and tubers.
Larvae length 2.5–3.5 cm
Family Tenebrionidae

253 Compost beetle
Chlaenius flaviguttatus
Several species of beetle are found in compost and living in the soil. This species is predatory and it feeds on other small animals. It is one of the bombadier beetles – see its defence mechanisms mentioned in Chapter 3.
Adult length 14 mm
Family Carabidae

COCKROACHES
See Chapters 2 and 3, also Illustration 200

254

254 Compost cockroach
This common, compost-dwelling native species mainly
eats decaying vegetable matter, but it has been suspected
of chewing roots and young shoots.
Adult length 20 mm
Order Blattodea

CRICKETS
See Chapters 2 and 3, also Illustration 197

255 Field cricket
Teleogryllus sp.
Field crickets are most noted for their 'evening song', made by rubbing wing parts together. They live within grass and plant litter, and are recorded as pests of potatoes.
Adult length 3 cm
Family Gryllidae

256 Mole cricket
Gryllotalpa sp.
Mole crickets, as their common name would indicate, spend most of their lives tunnelling in the soil. In rainy weather, they may leave the soil and become more conspicuous. They occasionally become pests of plant roots and may remove seed that has been planted. Their earthworks sometimes annoy fastidious lawn-keepers and bowling-green keepers.
Adult length 3.5 cm
Family Gryllotalpidae

CICADAS
See Chapters 2 and 3, also Illustrations 193, 194, 215, 329

257 Cicada nymph
Cicadas are associated with twigs and small stems.
However, the peculiar subterranean nymphs live deep in
the soil where they suck from roots. Sometimes, they are
dug up in the garden. This one is obviously a final-stage
nymph making its way to the surface to complete its life
cycle.
Nymph length to 2.5 cm
Family Cicadidae

SPRINGTAILS
See Chapters 2 and 3

258 Springtails
These microscopic insects are best known as quick-jumping specks that can occasionally be seen on seedlings, compost and fungi. They are common wherever there is decaying vegetable or animal matter, and sometimes they are minor pests of roots, seedlings and mushrooms.
Adult length 0.5 mm
Order Collembola

SPIDERS
See Chapters 2 and 3, also Illustrations 167 to 172, 230

259

260

259 Funnelweb spider
These large and dangerous spiders are more common and widely distributed along the east of Australia than generally realised because they remain hidden under rocks and logs and are active at night. If bitten, seek immediate medical help.
Adult size 4–6 cm
Family Hexathelidae

260 Wolf spider
These ground-dwelling spiders live within compost and leaf litter and prey actively on insects and other small animals. Their eyes are brightly reflective to torchlight at night.
Adult size 4–6 cm
Family Lycosidae

262 Redback spider
Latrodectus hasseltii
These highly dangerous, venomous spiders are widely distributed and frequently found among litter such as tins and bits of discarded metal close to the ground. The egg sacs are distinctive. If bitten, seek immediate medical help.
Adult size 2–3 cm
Family Theridiidae

261 Net-casting spider
Deinopis sp.
This fascinating spider rests motionless on tree trunks or other objects near the ground during the day, but in the early evening it begins spinning its special fishing net. By ~9pm it will be found poised over its trip lines with the net held ready to cast over any approaching prey.
Adult length 4 cm
Family Deinopidae

MITES

See Chapters 2 and 3, also Illustrations 11, 53 to 59, 195, 196, 212

263

263 Harvest mite
Erythraeus sp.
This very large mite is sometimes seen in compost and leaf litter. It is quite harmless to plants and feeds on decomposing vegetable matter and sometimes other small animals.
Adult length (with legs) 4 mm
Family Erythraeidae

CRUSTACEAE
See Chapter 2

264 Amphipod
This tiny, active, hopping animal looks something like a large flea. It is not an insect, as it is related to crabs and prawns. Amphipods play a very important part in the breakdown of plant litter to form soil and they are common in compost heaps.
Adult length 3 mm
Family Talitridae

265 Slater
These strange little animals are, along with amphipods, related to crabs and prawns. They live in compost heaps and under pots and stones, and they play a part in decomposition of plant remains. On rare occasions they have been suspected of damaging seedlings.
Adult length 6 mm
Order Isopoda

THRIPS
See Chapters 2 and 3, also Illustrations 60 to 64

266

266 Giant thrips
Idolothrips spectrum
This is one of the largest of all thrips; fortunately it is a
detritus feeder and does not harm plants. Sometimes it
can be found among fallen eucalypt leaves.
Adult length 12 mm
Family Phlaeothripidae

CHILOPODS AND DIPLOPODS
See Chapters 2 and 3

267 Centipede
Several species of centipede may be found in gardens. This species lives in compost heaps where it feeds on decaying vegetable matter and other small animals.
Adult length up to 17 cm
Family Scolopendridae

268 Long-legged centipede
This harmless animal lives in compost or under pieces of wood or other objects. It feeds on decaying vegetable matter, and is possibly introduced from overseas.
Adult length up to 5.5 cm
Family Scutigeridae

269 Millipede
(a) This harmless animal commonly inhabits compost heaps and other places where it finds shelter and decaying organic matter for food.
(b) This larger species of millipede comes from a drier environment.
Class Diplopoda

270(a)

270(b)

270(c)

270 Pill millipede
(a) Usually called pill bugs, pill millipedes are not related to true bugs. This is an adult rolled into a hard tight 'pill'.
(b) As the animal unrolls it exposes its legs.
(c) The heavily armoured animal trundles along like a tank.
Adult length up to 4 cm
Class Diplopoda

EARTHWORMS
See Chapter 2

271 Earthworm
Earthworms are usually highly beneficial animals because they dispose of leaf litter, recycle plant nutrients and help to aerate the soil. However, they can be serious pests of potted plants and are sometimes harmful in other situations. This common worm is not a native species.
Adult length up to 20 cm
Class Oligochaeta

SLUGS AND SNAILS
See Chapters 2 and 3, also Illustrations 201, 202, 248 to 250

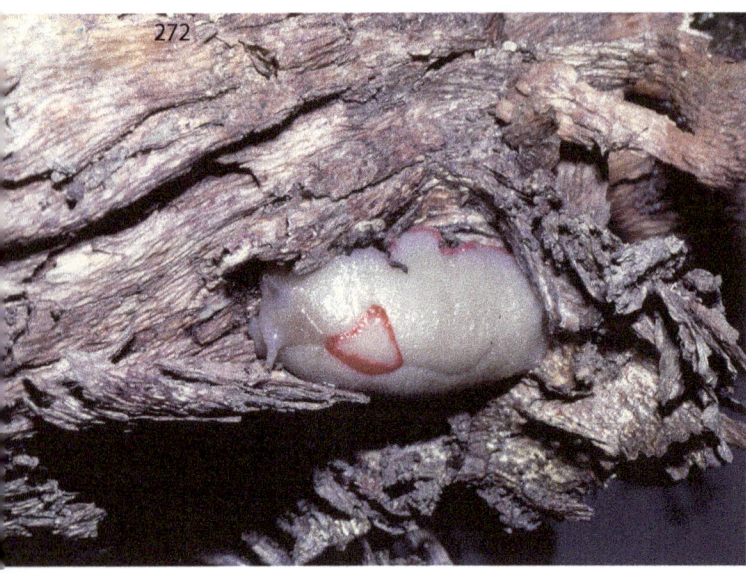

272

272 Red triangle slug
Triboniophorus graeffei
This native slug has a distinctive red, triangular marking. It is occasionally found in gardens close to bushland, but it is not regarded as a pest because it feeds on algae.
Length 6–9 cm
Family Athoracophoridae

REPTILES
See also Illustrations 233, 234, 235, 275, 276, 277

273

274

273 Legless lizard
Anomalopus verreauxii
Often mistaken for a snake and unfortunately killed as such, this handsome lizard can be identified by its tiny legs. It grows to 30 cm in length and it burrows in compost and loose soil, feeding on insects and other small animals. It is completely harmless, inoffensive and quite beneficial.
Adult length up to 35 cm
Family Scincidae

274 Red-naped snake
Furina diadema
Snakes usually get short shrift in gardens because many people have unreasoning dread of all snakes. However, this is one of several small harmless species that occasionally can be found in gardens. It averages about 25–30 cm in length but occasionally grows to about 50 cm. It is quite beneficial because it feeds on ants and other small animals.
Adult length up to 50 cm
Family Elapidae

FREE-RANGING SMALL ANIMALS

Many animals to be seen in gardens live a free life, ranging over an area of several suburban allotments or sometimes larger areas. Unlike the majority of small, garden animals, they are not restricted to specific food plants. Dragonflies, for instance, spend their immature adult stage far from the creek or dam where their nymphal stages live, then return to water for breeding once they have acquired full colour.

For further information relating to the following illustrations, the reader should refer to the appropriate section in Chapters 2 and 3.

REPTILES
See also Illustrations 233 to 235, 273, 274

275 Bearded dragon
Pogona barbata
In spite of their ferocious appearance, bearded dragons are very docile lizards that quickly learn to accept grasshoppers offered by hand. Besides grasshoppers, they eat beetles, other insects, snails and yellow flowers, and sometimes they graze on lawn grass. Sometimes they can be observed climbing over leaves at the top of eucalypt trees, eating Christmas beetles. Bearded dragons are often wrongly called frilled lizards and, unfortunately, these and other lizards are soon killed off by household pets.
Family Agamidae

276 Jacky dragon
Amphibolurus sp.
This dragon became domesticated, following people around the garden to look in every hole being dug. It also would sit on the dog, with the dog's permission, and catch flies.
Family Agamidae

277 Skink
Several species of lizard live in gardens and all are harmless and highly beneficial because they eat moths, beetles, snails and many other pests. Unfortunately, they are highly vulnerable to cats and dogs. This small ground-dwelling species shelters and feeds amongst sprawling plants, leaf litter, stones, pots and other objects.
Family Scincidae

WASPS
See Chapters 2 and 3, also Illustrations 3 to 5, 110, 143 to 153

278 Mud-dauber wasp
Sceliphron sp.
(a) This is one of many species that gather clay to construct nests of several cells. Each cell is provisioned with paralysed spiders, caterpillars or other insects. The wasp lays an egg in each cell and the resulting maggot consumes the provisions and pupates within the cell.
Adult length 2.5 cm
Family Sphecidae
(b) The contents of another mud-dauber wasp nest showing the paralysed caterpillars this species stores as food for its larvae. This group of wasps is of great benefit in the garden. The victims are paralysed rather than killed so that they will not decay before the wasp larvae devour them.
Family Sphecidae

279

280

279 Mud wasp, key hole wasp
Pseudabispa bicolor
Sometimes this is called a key hole wasp because of its readiness to block key holes and similar openings with mud cells and paralysed prey.
Adult length 1.4 cm
Family Vespidae

280 Mud wasp nest
Some species of mud-dauber wasps construct their mud cells on leaves or twigs of plants as shown in this illustration.

281(a)

281(b)

281(c)

281(d)

281 Burrowing wasp

Cryptocheilus sp.

(a) Burrowing wasps are busy hunters, seeking out suitable prey – often spiders. The paralysed victim may be dragged for a considerable distance to where the wasp has selected a suitable site to excavate a tunnel.

(b) The wasp works hard excavating and enlarging the tunnel to accommodate the size of the prey.

(c) When the tunnel is large enough, and after several tries, the victim is dragged below ground where the wasp deposits an egg or eggs on the prey before carefully filling the hole.

(d) The wasp gradually fills the hole and, using its body, it tamps the soil down firmly, finally disguising the surface with twigs or small stones. In due course, its progeny dig their way to the surface as adult wasps.

Adult length 3.5 cm

Family Pompilidae

282 Wax wasp
Instead of using mud to construct its cells, this unidentified wasp collects the resin from trees.
Adult length 2–3 cm

MOTHS AND BUTTERFLIES
See Chapters 2 and 3, also Illustrations 65 to 81, 109, 114, 115, 122 to 136, 221 to 226, 245

283 Procession caterpillar

(a) These caterpillars and their moths consist of several species with viciously stinging hairs that cause serious allergic reactions. They trek long distances, the caterpillars touching end to end.

(b) The eggs are laid by the moth on the base of a tree – usually a wattle. People have had to seek medical help after touching the egg mass.

(c) A defoliated shrub after the feeding of these caterpillars.

(d) The bag around the base of a tree where the caterpillars shelter during the day. The debris in the bag continues to be a hazard long after the caterpillars have gone. In inland districts the caterpillars breed on larger trees and construct hanging shelter bags. The dust from these bags has hospitalised children.

Adult length up to 5 cm

Family Notodontidae

DRAGONFLIES AND DAMSELFLIES
See Chapters 2 and 3

284 Dragonfly

(a) Dragonflies are usually associated with pools of water in which they breed. Their nymphs are aquatic. It is a surprise, therefore, that they are common even in gardens a considerable distance from water. When the adults first emerge from their aquatic stage, they usually move away from the breeding area while they develop full, mature colouring. It is during this time that they are seen in gardens. They seize and eat other flying insects and so they are beneficial.

(b) The larval stage is aquatic and preys on other small aquatic creatures.

Wingspan 6 cm

Suborder Anisoptera

285 Damselfly

Like the dragonflies, damselflies have aquatic larvae and the adults must return to water for breeding. They also capture and eat other insects. Damselflies can be distinguished from dragonflies by the way the wings are folded back down the body while the insect is at rest.

Adult length 3.5 cm

Suborder Zygoptera

FLIES

See Chapters 2 and 3, also Illustrations 1, 6, 116 to 121, 139 to 142, 328

286 Robber fly
Neoaratus sp.
This large, aggressive fly captures other insects, often on the wing. Its saliva quickly immobilises the victim and enables the fly to suck out the contents.
Adult length 2.7 cm
Family Asilidae

287 Crane fly
Ptilogyna ramicornis
This large, ungainly and somewhat fragile insect loses its brittle legs if caught or handled. It probably feeds on nectar.
Wingspan 3.5 cm
Family Tipulidae

BUGS
See Chapters 2 and 3, also Illustrations 41 to 52, 173 to 176, 227, 228

288

288 Giant water bug
Lethocerus insulanus
These large aquatic bugs capture and suck dry other
aquatic creatures such as fish and frogs. They can fly and
in rainy weather sometimes appear in gardens.
Adult length 5–7 cm
Family Belostomatidae

AMPHIBIANS

289 Frog
Litoria sp.
This small tree frog shelters among leaves and gives a sharp call when rain falls or sprinklers operate.
Family Hylidae

290 Frog
Litoria caerulea
Although very sleepy and inactive during the daylight, frogs are very active nocturnal animals that devour large numbers of pests, particularly egg-laying adults.
Family Hylidae

BIRDS

291 Black cockatoo damage
Yellow-tailed black cockatoos hunt for borers in trees and, after locating them by listening, they tear branches and trunks apart to access the grubs.

292 Frogmouth
Podargus strigoides
These birds are more common than realised because they sit disguised and motionless during the day.
Family Podargidae

293 White-cheeked honeyeater
Phylidonyris sp.
Grow the right plants and discourage currawongs, crows, magpies, butcher birds, kookaburras and noisy miners and your garden will be a home for countless small birds.
Family Meliphagidae

MAMMALS

294 Echidna
Tachyglossus aculeatus
Gardens near bushland or in the country are often visited by echidnas. They usually stay around for a few weeks, active at night and hidden during the day.
Family Tachyglossidae

295 Sugar glider
Petaurus breviceps
Although this beautiful little marsupial is commonly called the sugar glider, its main food is insects. In particular, it relishes leaf-eating beetles, grasshoppers, moths, caterpillars and even young birds. Their role in home garden pest control (pests such as Christmas beetles) could be immensely important. Unfortunately, along with other potentially important pest-eating native animals and birds, they are vulnerable to cats and dogs.
Family Petauridae

DISEASES AFFECTING GARDEN PLANTS

Garden plants are attacked by several different disease organisms. Most of those recorded in gardens are fungal and often they indicate plants that are being grown in an unsuitable environment. Plants from more temperate climates and the winter rainfall zone, for instance, are often plagued by fungal diseases when grown in tropical, high humidity, summer rainfall zones. Similar problems emerge when tropical plants are grown in a Mediterranean climate.

Plants that are continually attacked by disease should be replaced by species that are less susceptible. Many native and other plants have few disease problems if wisely selected for the locality.

For further information relating to the following illustrations, the reader should refer to the appropriate section in Chapters 2, 4 and 5, and Appendix I.

FUNGUS
See Chapters 2 and 4, also Illustrations 326 to 332

296 Cylindrocladium leaf spot
Cylindrocladium sp.
This severe fungal leaf spot attacks melaleucas and related plants, causing necrotic or dead areas often surrounded by a purple margin. Affected leaves tend to fall readily from the plant.

297 Verrucispora leaf spot
Verrucispora proteacearum
This severe leaf-spotting disease is caused by a fungus. Grevilleas and hakeas are commonly attacked and infections are often worse in wet weather.

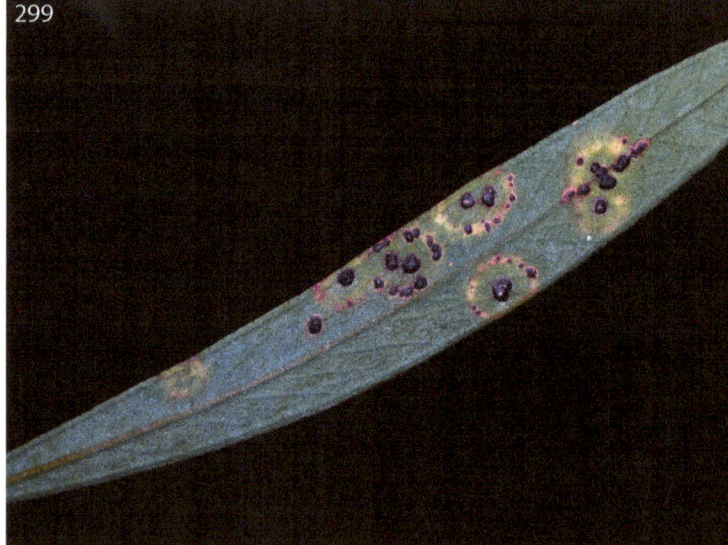

298 Sooty spot
Placoasterella sp.
This superficial fungal spot does not affect the underlying plant tissue and can, with difficulty, be wiped off. It mainly affects grevilleas and hakeas.

299 Tar spot
Phyllachora sp.
This fungus is often mistaken for scale insects and it is sprayed as such quite ineffectively. The small, hard black, irregularly shaped fungal bodies are often arranged in a more or less circular pattern. They cause yellowing and leaf fall. Callistemons are the most common host plants; mainly *C. Gawler* hybrid and *C. speciosus*.

300 Powdery mildew
Oidium sp.
(a) This fungus attacks a wide range of plants, and damage varies from fairly superficial whitish mouldy patches on leaves to spotting and dead patches on the surface of leaves. Plants from dry climates are susceptible when grown in humid, coastal districts.
(b) Close inspection shows the delicate white mouldy growth.

301 Eucalyptus leaf spot
Phaeoseptoria eucalypti
This fungal leaf spot commonly infests the foliage of eucalypts from arid environments when they are grown in wet, humid, coastal districts.

302 Acacia rust galls
Uromycladium sp.
(a), (b) Acacias are the main host for this group of fungi, which cause grotesque galls on young shoots and leaves. Often insects and mites live and feed in or on the galls. Heavily infected trees are severely disfigured. Early removal by pruning may help reduce infections.

303 Leaf and stem rust
Uromyces sp.
Rust is a name often wrongly applied to any rusty discolouration of leaves or stems. Rust fungal diseases are characterised by raised brown or yellow pustules on the surface of leaves or young stems. Affected parts become deformed and unsightly. The pustules break open to release the rust-coloured spores, as illustrated.

304 Myrtle rust
(a) Fruit of the exotic jaboticarba showing the yellow spores of myrtle rust. Rust diseases are usually very specific to just one family of plants – myrtle rust to only plants of the family Myrtaceae. Chemicals are available to manage myrtle rust infections. See the reference to chemical registrations in Chapter 5.
(b) The severe effect on *Xanthostemon youngii* is shown. This disease, accidentally introduced to Australia, has the potential to have a serious impact on Australia's dominant myrtaceous flora.

305 Ramularia blight
Quambalaria sp.
Eucalyptus citriodora is the main host for this fungus. The young shoots and leaves look as though they have been splashed with white paint. Affected leaves and shoots are severely distorted.

306 Shoot blight pestalotiopsis
Pestalotiopsis sp.
Young leafy shoots are the usual target for this fungus. Outbreaks may be associated with too much shade and lack of air movement. The cultivars of *Melaleuca bracteata* are sometimes affected by this disease.

307 Shoot blight alternaria
Alternaria sp.
This disease may appear under wet conditions on sites where the number of full sunlight hours is short.

308 Banksia canker
Pestalotia sp.
This fungus causes irregular swellings with sunken sections on the branches. The branch above the infection dies and the plant shoots temporarily below the canker. Pruning well below infected sections of stems and treatment of cuts with fungicide may help reduce infections.

309 Bracket fungus
Frequently, wood-rotting fungi, such as the bracket fungus illustrated, gain entry to the branches or trunks of trees through old pruning scars. Pruning cuts should be sealed with a liquid grafting compound.

310 Phellineus
Phellinius noxius
This very serious fungal disease is expressed as a brown fungal growth that extends from the soil and roots up the lower trunk and causes the death of the upper portions of affected trees.

311 Crown rot sclerotium
Sclerotium rolfsii
(a) Soft-stemmed plants such as annuals and herbaceous perennials may be attacked by this fungus. Plants with a dense cover of leaves close around the crown, such as the *Xerochrysum* illustrated, are very susceptible under damp conditions.
(b) As the woolly fungal growth and dead plant dry out, the fungus forms sclerotes or resting bodies which look like radish seeds. Infected plant material should be burnt and not put into the compost heap where the sclerotes would germinate.

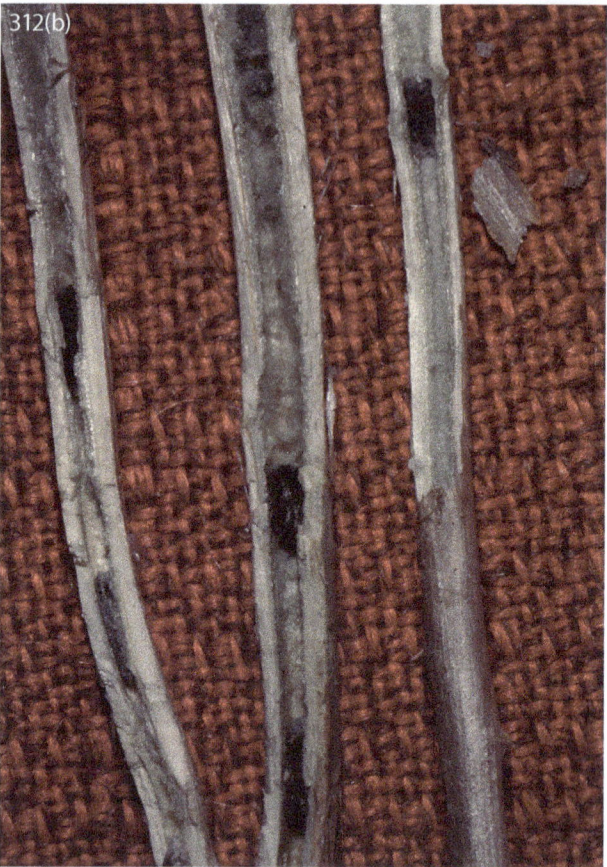

312 Stem rot sclerotinia
Sclerotinia sclerotiorum
(a) This disease affects soft-stemmed plants (such as stocks and rosellas), causing sections of stems to dry out.
(b) If dried stems are split open, the black sclerotes or resting bodies may be seen in the pithy centres of stems.

313 Pink disease
This disease appears in long wet season and high rainfall areas. It causes the death of branches and shows on the bark as pink spots. Citrus trees and other trees are attacked.

PROTISTS OR PROTISTA
See Chapters 2 and 4, also Illustration 325. Phytophthora is no longer classed as a fungus

314 Collar rot
Phytophthora sp.
This disease becomes active under wet or waterlogged conditions and attacks roots and below-ground stems. It kills the bark and ringbarks plants, which quickly die.

315 Stem-gumming phytophthora
Phytophthora parasitica
(a) The early symptom of this disease is 'gumming'. A close examination will show that the gum is exuding from splits in the bark.
(b) Dead roots will also be found and the fungus may extend as a collar rot.

BACTERIAL DISEASES
See Chapters 2 and 4, also Illustration 333

316 Iris bacterial leaf spot
Bacterial spots frequently have a water-soaked appearance and affected leaves tend to disintegrate rapidly.

317 Bacterial spot
This common bacterial spot on *Brunsfelsia* is a dry spot similar to a rusty red bacterial spot on bougainvillea.

318 Crown gall
Agrobacterium tumefaciens
This bacteria causes a large, soft gall to develop at the base of affected plants.

VIRAL AND VIRUS-LIKE DISEASES
See text Chapter 4, also Illustration 334

319 Rose mosaic virus
The lacy, chlorotic pattern indicates this rose bush is infected with rose mosaic virus. This disease can be spread by contact or on secateurs and other garden tools. Incineration is the only treatment.

320 Dahlia ring spot
The distinctive concentric pattern indicates this plant is infected with ring-spot virus. It should be removed and hands and tools thoroughly cleaned. Incineration is the only treatment.

321 Calico virus
This virus disease is sometimes seen on weeds on farm headlands and garden surrounds from where it may be spread by insect or tools. Affected weeds should be removed and others controlled.

322 Orchid ring-spot virus
Pale green or yellow ring-shaped leaf spots are symptoms of infection by this virus. Infected plants slowly decline.

323 Orchid colour break

(a) This virus is expressed as colour break in flowers and mild deformity of petals.

(b) The leaves of orchids infected with colour break may show sunken rings or streaks.

(c) A colour break-infected flower showing deformity.

(d) Virus symptoms expressed in orchid leaves.

324(a)

324(b)

324(c)

324 Bigbud or virescence

(a) Green flowers and grotesque floral deformities are caused by this virus-like organism infecting this flat weed. Tomatoes are often infected.

(b) Normal flowers are replaced by tufts of deformed flowers emanating from affected flower heads.

(c) A phlox plant showing the typical deformed clusters of green flowers.

HARMLESS AND BENEFICIAL OTHER ORGANISMS AND SPECIALISED ROOTS AND BARK

As with insects and other small animals, a significant number of fungi, protists, bacteria and virus-like organisms are either harmless or directly or indirectly beneficial. Illustrated is a range of these organisms and expressions of their activity.

For further information relating to the following illustrations, the reader should refer to the appropriate section in Chapter 2.

PROTISTA OR PROTISTS
See Chapter 2 and 4, also Illustrations 314, 315

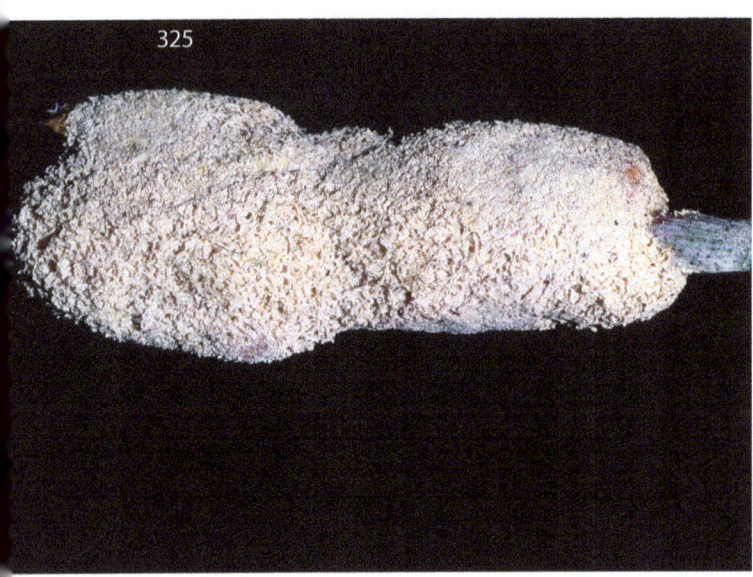

325

325 Slime mould
Myxomycetes
This very primitive slime is now classified as belonging to the kingdom Protista and is no longer considered a fungus. It may be relatively inconspicuous in its early stage, moving over the surface of decaying vegetable matter. When it is ready to spore, it creeps up nearby blades of grass or other objects and turns into a frothy mass of grey, mustard-coloured or pinkish spores. It is harmless but somewhat repulsive. If considered necessary, the mass of spores may be brushed or hosed away or sprayed with a fungicide such as copper oxychloride.

FUNGUS
See Chapters 2 and 4, also Illustrations 296 to 313

326(a)

326(b)

326(c)

326 (see also next page) Stinkhorn fungus
(a) The many species of stinkhorn fungus have come into prominence since mulching of garden beds has been promoted. They feed on decaying vegetable matter – mulch. Most give off a most unpleasant smell when they reach maturity; otherwise they are harmless.
(b), (c), (d), (e) The species vary considerably in appearance, but the fruiting body comes from an egg-like ball produced among the white thread-like mycelium.

326(d)

326(e)

327

327 Unidentified stinkhorn fungus
A basket-like species of stinkhorn that is not often seen.

328(a)

328(b)

328(c)

328 Stinkhorn fungus

(a) This red, octopus-like stinkhorn shows the brown, smelly spore mass.

(b) This similar species shows the mass of flies that are attracted to the smelly spores.

(c) A fruiting body that has been licked clean by the flies that spread the spores.

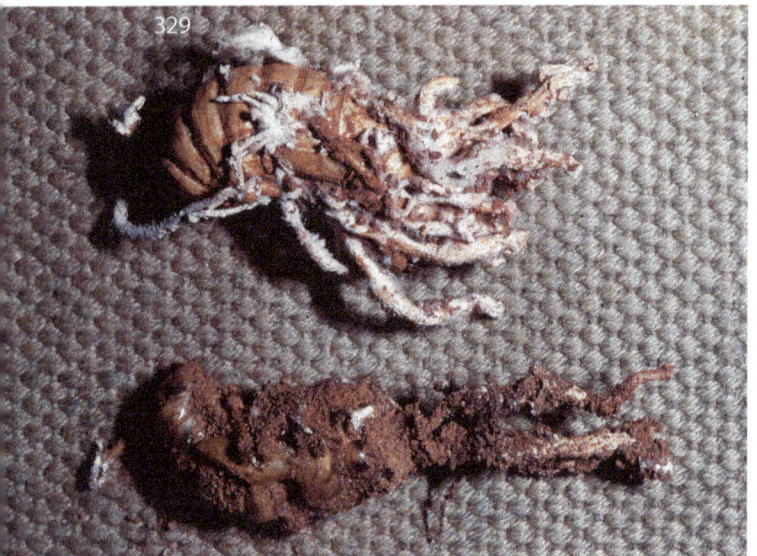

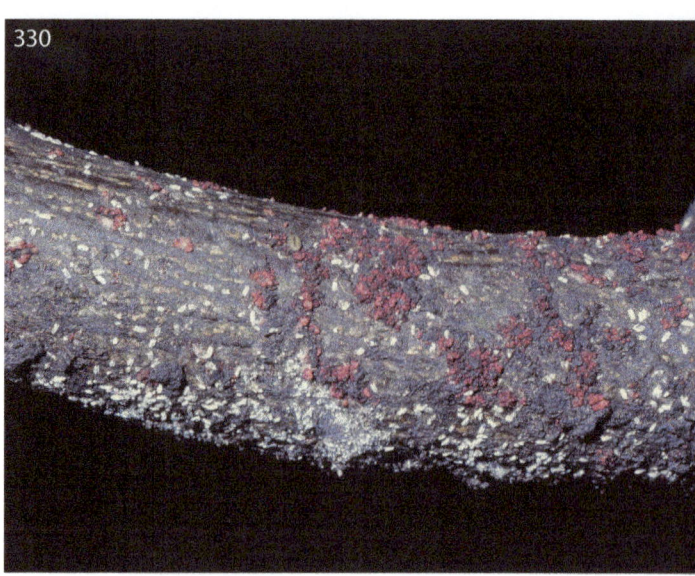

329 Cicada nymph fungus
Cordiceps sp.
This soil-dwelling fungus attacks insects and other small animals living in the soil. These cicada nymphs are an example and show the fruiting bodies of the fungus extending from the victims.

330 Entomogenous fungus
The red particles of one of the entomogenous fungi can sometimes be seen on the trunks of old, neglected fruit trees. This species, shown here with white louse scale, is a very useful parasite of scale insects.

332 Mycorrhiza
These growths on roots are an example of a symbiotic relationship between a plant and a fungus. They mostly occur on the roots of plants that grow on very nutrient-deficient soils and help the plant to access nutrients in exchange for water and other elements.

331 Entomogenous fungus
This orange entomogenous fungus quite effectively parasitises and controls scale insects.

BACTERIA
See Chapters 2 and 4, also Illustrations 316 to 318

333 Rhizobium

As distinct from the root galls caused by root knot nematode [Illustration 247], these nodules are attached to the roots – not within the roots. They are formed by beneficial bacteria that live in a symbiotic relationship with some plants. The bacteria access nitrogen from the soil air for the plant in exchange for water and other elements.

VIRUS
See Chapter 4, also Illustrations 319 to 324

334 Caterpillar virus
This caterpillar is infected with a virus disease. This virus is one of several diseases that occasionally play an important part in control of caterpillars.

LICHENS
See Chapters 2 and 4

335 Lichen
(a) Lichens are lowly forms of life consisting of a fungus and an alga combined in a close association. They obtain their food from the air and accumulated plant debris. They do not harm plants, but their presence indicates neglected or sick plants. This photograph shows a range of different lichens on a dead branch.
(b) Lichen growing on a fence post

OTHER PLANTS AND ODDITIES

336 Proteoid roots
(a), (b) These peculiar clusters of hairy roots are produced on many proteaceous plants such as grevilleas, hakeas and banksias as an aid to accessing nutrients from very low-nutrient soils.

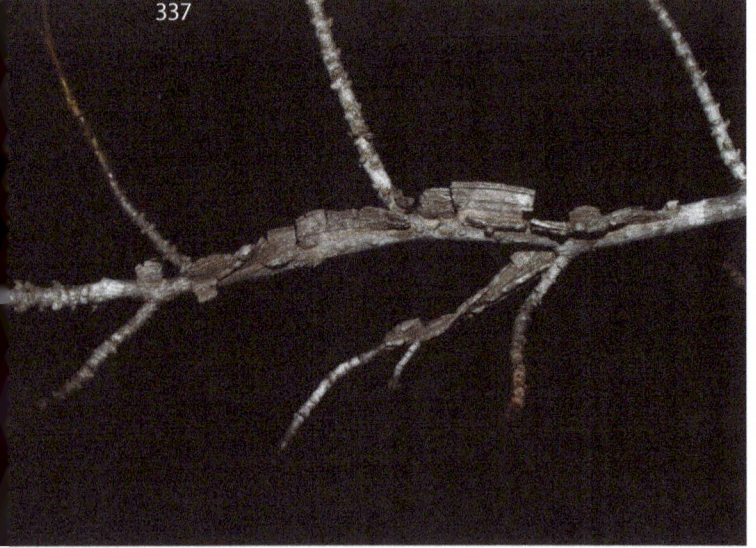

337 Bark cork – liquidambar
Over the years garden complaints have included a 'disease' affecting liquidambar trees. This 'disease' has persisted in spite of sprays and pruning. The corky flanges on twigs are a natural feature of a few plants and supposedly add character to the plants while deciduous.

Illustrations X

PARASITIC PLANTS

Besides parasitic fungi, bacteria and animals, there are some plants that parasitise other plants. Some parasitic plants attach to and live on the above-ground parts of their host (for example, Australian mistletoe and dodder), while others have roots that attach underground to the roots of their host plant. These latter include the spectacular West Australian Christmas tree (*Nuytsia floribunda*) and the exotic *Rafflesia arnoldi*, which bears the largest single flower in the world – up to one metre in diameter. Some of these entities are only partially parasitic because they have chlorophyll and can synthesise some of their own nutrients, but take water and soil-derived elements from their host. Others such as the *Balanophora* [Illustration 345] have no chlorophyll and are therefore total parasites.

For further information relating to the following illustrations, the reader should refer to the appropriate section in Chapter 2.

338 Cuscuta dodder

Cuscuta sp.

(a) The seeds of this leafless parasitic plant germinate in the soil in the normal way. The twining stems quickly grasp nearby plants of any kind and the haustorial suckers of the dodder penetrate the surface of the host plant and absorb its nutrients. The root of the dodder soon withers and the upper parasitic portion may extend vigorously over whatever plant it contacts.

(b) The trailing stems of cuscuta are usually orange in colour. Note the haustoria penetrating the leaf surface. Family Convolvulaceae

339 Laurel dodder

Cassytha sp.

(a) Laurel dodder is much more vigorous than the cuscuta dodder and grows high up in trees. The stems are green or purplish brown.

(b) Stems twine and attach in the same way as the cuscuta dodder, but the fruit are quite distinct. Family Lauraceae

340 Mistletoe running roots
This type of mistletoe has clinging roots that extend along the branch and penetrate the surface with haustoria every few centimetres.

341 Mistletoe tuber
Other species of mistletoe form a woody crown at the point where they first attach and have no obvious running roots.

342 Mistletoe flowers
Amylotheca dictyophleba
Some mistletoes have beautiful flowers and are worthy of horticultural attention.
Family Loranthaceae

343 Mistletoe seeds germinating
The seeds of mistletoe are coated with sweet-tasting, very sticky mucilage. Birds feed on this sugary substance and wipe the seeds onto nearby twigs and branches or even fencing wire with their rear end. Mistletoe seed germinates where it is deposited, and the root quickly penetrates the bark of host plants.

344

345

345 Balanophora
Balanophora fungosa
This plant mimics a fungus. It attaches to the roots of trees
and takes nutrients from the host plant.
Family Balanophoraceae

344 Secondary mistletoe
Notothixos sp.
Some species of mistletoes infest only other species of
mistletoe. This one has almost killed the primary species.
Primary and secondary mistletoes can also be found on
root parasitic plants such as *Exocarpos*.
Family Viscaceae

PHYSIOLOGICAL DISORDERS

While plants suffering a physiological disorder may appear to be diseased, no pest or pathogen (insect, mite, fungi, bacteria, or virus) is present. The abnormality may be caused by environmental factors such as: climatic conditions significantly different from where the plant originates; salty soil or water or salt-contaminated fertiliser; extreme soil pH conditions causing excess or deficiency of trace elements; or waterlogging of soil. Climatic factors may be to do with day length or strength of sunlight, maximum or minimum daily temperature, humidity, soil pH or soil moisture/air balance.

Symptoms include: dead sections or discolouration of leaves; periodic falling of leaves, flowers or fruit; dieback; or fruit splitting. Treatment involves attempting to modify the conditions to better suit the needs of the plant or to replace it with another that is more suited to the conditions.

346 Fasciation
This abnormal, strap-like growth, illustrated here on a callistemon, may occur on any type of plant. It is a physiological condition brought about by unfavourable growing conditions. It occurs most frequently in plants that are being grown in conditions far different from where the plant grows naturally.

347 Air pollution
If you have ever ridden for some distance in the back of a truck or ute, you will have been subjected to excessive carbon monoxide. This plant was part of several consignments that arrived in good condition after a 100 km trip but in the next few days developed this condition from carbon monoxide poisoning.

348 Sulphur dioxide pollution
This occurred in a mining town where sulphur dioxide was present in the emitted pollutants. If the wind changed to over the town, an emergency control shut the system down for safety. This garden was nearest the mine and the gardener was watering plants when the wind shift occurred. Sulphur dioxide plus water makes sulphuric acid.

349(a)

349(b)

349(c)

349 Waterlogging

(a) Dead palms at the setting up of Expo 88. The drainage system did not function.

(b) The holes from which the drowning palms had been removed.

(c) Another tree with water visible at the surface

350

350 Freezing

The effect of below-freezing temperatures on a sensitive plant. Contrary to some promoted ideas, a 'warm' mulch around plants for winter actually increases the formation of frost because it insulates radiation from the soil.

TOXICITY
Symptoms can occur from contaminants such as salt in soil, water or fertiliser or from oversupply of nutrient elements

351
(a) Salinity – water
This photograph was taken in a town where the water supply contained too high a level of salt for a salt-sensitive plant such as a rose.
(b) Salinity – drought
This town's water supply turns salty in drought times as the photo shows.

352 Salinity – fertiliser
(a) This photo was taken after a statistically arranged trial comparing fertilisers. This was an organically approved fertiliser stating on the bag 'Totally organic – cannot burn plants'. In fact, any fertiliser based on chook manure will be very salty.
(b) Illustrating the effect of salty fertiliser on affected foliage

353

(a) Boron toxicity – pavers
This occurred at a racecourse where the management paved a much worn area under trees. Usually, before laying pavers, operators treated the area with a borax preparation to prevent weeds. This was the effect of the ensuing toxicity.

(b) Boron toxicity – wood shavings
This gardener was quite emphatic that no boron had ever been used, but long questioning revealed his garden had been mulched with shavings and sawdust from a wood turner. Some logs are treated with borax to prevent borers.

354 Fluoride toxicity
This occurred in a garden where an unused bore was brought into use without knowing the bore water contained very high levels of fluoride.

356 Phosphorus toxicity
Several Australian native plants from very infertile acid soils low in phosphorus are intolerant of fertilisers containing phosphorus. The worst affected are banksias, hakeas, grevilleas and related plants. Symptoms include stunted growth, yellow leaves and dieback of leaf and shoot tips. The plant on the left was given no phosphorus, while the plant on right had a normal application of fertiliser containing phosphorus. The symptoms resemble lime chlorosis, but there is little response to iron chelate.

355 Manganese toxicity
If the pH of a soil is below 5.0 (very acidic) and the soil contains relatively high amounts of manganese, this purplish toxicity symptom may occur because the strongly acidic condition makes manganese more soluble.

DEFICIENCY

Symptoms result from shortage of an element or from too much of one element inducing a shortage of another

357 Boron deficiency
(a) Shortage of boron causes abnormalities such as, in citrus, lopsided dry fruit and, in palms, stunted leaves with dead tips and dry black scarring.
(b) Corky dry black scarring on young leaves, stems and growing points from lack of boron.

358 Zinc deficiency is expressed as small leaves with yellow mottling between the veins, bunched close together (rosetting).

359 Magnesium deficiency
(a) This deficiency affects the old leaves and is expressed as a green triangular base to the leaves with yellow tips and edges or as shown here on a pinnate leaf.
(b) Besides possibly being a shortage of magnesium, deficiency symptoms can be induced by over-application of lime, either as lime or gypsum as illustrated here.

360 Calcium deficiency
Calcium deficiency is expressed as a dry black scarring on one side of stems causing bending and distortion, particularly to flower stems.

361 Phosphorus deficiency
Plants deficient in phosphorus are stunted and show reddish or purplish discolouration. It is most often seen on tomatoes and cabbage relatives.

362 Iron deficiency
Plants that grow in acidic soils have a high requirement for iron because iron is more soluble under acidic conditions. Applications of lime or dolomite make the soil less acidic, reducing the availability of iron, and may induce the iron deficiency symptoms of smaller yellow leaves.

363 High pH
This symptom of yellowish or reddish discolouration on an acid-loving plant results when the soil has a high pH – above 7.0. It is in fact a shortage of iron; phosphorus toxicity has similar results and may confuse the diagnosis.

Illustrations XII

HORTICULTURAL PROBLEMS

Some plant problems or abnormalities are simply caused by horticultural mismanagement – a procedure such as transplanting, fertilising, spraying or supporting that has not been carried out as carefully or thoughtfully as required to suit the needs of the plant.

A typical example is whipper snipper 'disease' of palms, whereby the supporting band of roots from the base of the stem is gradually cut away and the base undermined by the wrong use of this tool. In extreme cases the palm eventually falls over.

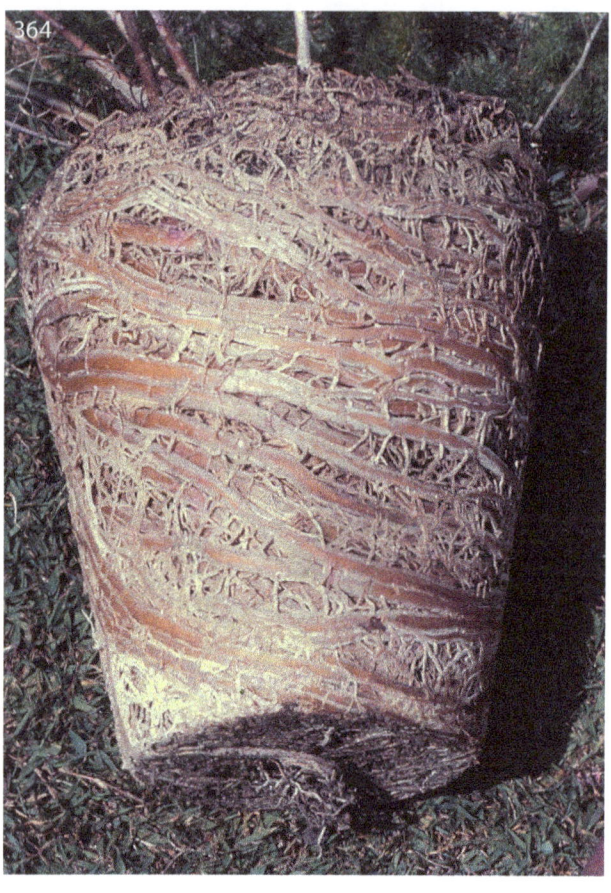

364 Root-bound plant
This is an example of a plant that has been held in a container for far too long. Some plants that have a particularly vigorous root system are more prone to develop this condition.

365 Transplant problem
This strange, double acute bending of roots occurred because the seedlings were well past the optimal stage for transplant. The plant had a long root system which was stuffed into a dibble hole, resulting in bent roots as the plant grew. A different transplanting technique and trimming off several centimetres of the root ends would have prevented the problem.

366 Planting-depth problem
The roots of plants require some air and the depth they grow to is in balance with the permeability of the soil in which they are planted. This requirement varies with species of plant – eucalypts are among the most sensitive and usually die about 12 months after a few inches of soil have been added over the existing soil level. Potted plants planted deeper than about 10 cm above the top of the pot level are likely to die within 12 months.

367 Cincturing
This is an extreme example of a plant that had a label tied on with a tie containing wire. The label disintegrated in time and the tie was overlooked. As the plant grew the tie cinctured the stem and became deeply imbedded.

368 Whipper snipper 'disease'
Misuse of this useful garden tool results in this undercut condition, which is commonly seen in public and private gardens on a variety of trees. Palms are particularly subject to this mismanagement and can be so severely damaged that they blow over.

369 Spray drift
(a) With the advent of selective herbicides for garden use on weeds in lawns the symptoms of herbicide damage from drift of invisible spray droplets have become commonplace. Another hazard is spray accidentally applied to surface roots of nearby trees.
(b) This example is typical of the symptoms of the drift of invisible spray droplets of Roundup™.

370 2,2-DPA

A local authority crew ran into this problem using 2,2-DPA along kerbs and footpath edges to control grass and weeds. In particular, palms, which have a shallow, widespread root system, were distorted and died.

371 Ronstar™ damage

Ronstar™ is a useful chemical for preventing weeds in potted plants, but it has to be applied carefully. In this case a few grains lodged in the top of young palms.

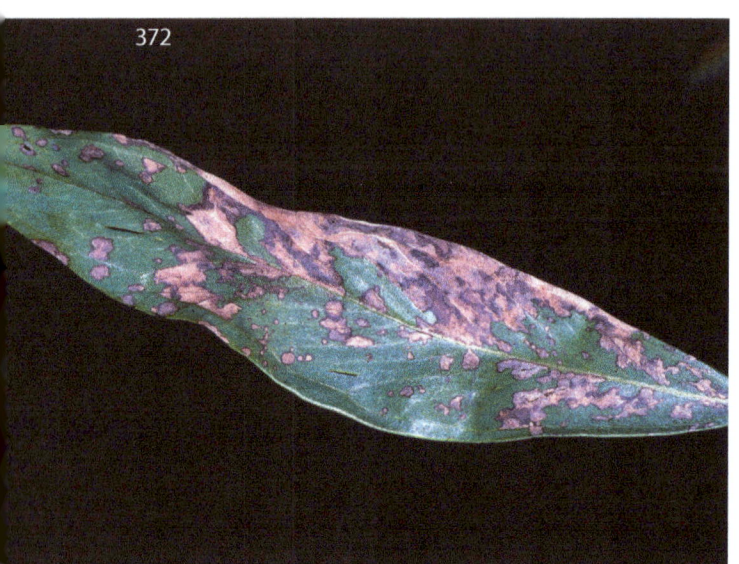

372 Spray burn

This gardener sprayed healthy plants as a precaution against disease. When this plant showed dead spots the plant was sprayed again at a stronger concentration – and then again. Next a complaint was lodged about the ineffectiveness of the chemical. The spots were actually a symptom of phytotoxicity caused by the spray on this spray-sensitive plant. Plants are like people – no matter what is applied, there will be at least one individual that reacts unfavourably to it.

373 Pot weeds

Weeds, particularly those that bear fruit to attract birds, can be spread far and wide. Every pot in this nursery, more than 100 m from wild tobacco trees, *Solanum mauritianum*, became infested with seedlings deposited by birds.

With so many pesticides on the market, it is now a real challenge for home gardeners and occasional users of pesticides to identify the safest and most effective product for their purpose. Pesticides must be used only as per the label on the product you purchase. Product labels are very specific about the circumstances in which that product can be used – for example, on what plants, which pests, and application rates. There may be dozens of different products containing the same active ingredient and each of those products will have different label instructions, especially where the products have different concentrations of active ingredients or mixtures of active ingredients. The product choice is made less complicated because retailers usually stock a limited number of products with the same active ingredients. Current registrations can be checked online in the PUBCRIS database on the Australian Pesticides and Veterinary Medicines Authority website, www.apvma.gov.au. It should be noted that registrations are continually changing. To assist in identifying a product that is registered for your purpose, the following table lists some of the commonly used active ingredients for pest groups.

Pest or disease	Treatment or active ingredient Note that not all pesticides containing these active ingredients will be registered for your specific purpose – always follow the product label
Chewing insect pests	
Caterpillars Grasshoppers Beetles	Synthetic pyrethroids such as tau-fluvalinate and cyfluthrin or pyrethrin Biological pesticides derived from *Bacillus thuringiensis* are effective against caterpillars only, but require a high population of caterpillars to be effective
Fruitfly Codling moth	Yeast-based baits with Spinosad™ – a product derived from soil bacteria
Leaf miners	Horticultural oils, such as pest oil, white oil, summer oil
Borers	Pruning, a wire probe and careful application of pyrethrin
Slugs or snails	Slug and snail baits of metaldehyde or methiocarb
Sucking insect pests	
Aphids Leafhoppers Whitefly	Imidacloprid, pyrethrin, Natrasoap™ products
Scales	Horticultural oils
Mealy bugs Plant bugs	Imidacloprid
Thrips	Imidacloprid, pyrethrin, Natrasoap products
Mites	Tau-fluvalinate, sulphur, lime sulphur products, Natrasoap™ products
Fungi	
Powdery mildew	Mancozeb plus wettable sulphur
Downy mildew	Myclobutanil, mono-dipotassium phosphite
Rust	Mixtures containing tebuconazole or trifloxystrobin
Leaf spots	Mancozeb plus wettable sulphur
Stem rots Crown rots Root rots	Mancozeb plus wettable sulphur Mono-dipotassium phosphite
Damping off	Mancozeb plus wettable sulphur
Bacteria	
Spots	Copper sprays such as copper oxychloride or cupric hydroxide
Wilts	Long crop rotation and use of disease-free planting material
Viruses	
	Early removal and incineration before the virus spreads
Nematodes	
Leaf nematodes	Pruning and drier conditions
Root nematodes	Maintaining a 120–150 mm layer of organic mulch is the best long-term remedy

Appendix II: Glossary

acaricide   preparation for eradicating mites

adult   final stage in the life cycle

asexual   without sex

biological control   control of a pest by a parasite or predator

blight   withering of shoots caused by a parasitic disease

canker   an area of dead tissue resulting from invasion by disease

chlorophyll   the light-absorbing green pigment of plants

cuticle   outer covering or skin

cytogenetics   the study of cell structure and organic variation

detritus   waste material or litter

ectoparasite   a parasite that feeds externally on its host

egg sac   sac-like structure produced to hold eggs

endoparasite   a parasite that feeds internally in its host

entomogenous   feeding on insects

excretion   cast-out waste matter

fission   division or splitting

fruiting body   sexually reproductive structure of a fungus

fungicide   preparation for eradicating fungi

gregarious   growing or living in a close community

hermaphrodite   organism with both male and female parts equally functional

honeydew   sugary solution excreted by some sap-sucking insects

hyperparasite   parasite of a parasite

insecticide   preparation for eradicating insects

larval stage   immature feeding stage of some insects

lesion   a wound or localised area of infection

metamorphosis   the change in an insect's form from larva to adult

miticide   preparation for eradicating mites

molluscacide   preparation for eradicating slugs and snails

nectar   sugary substance secreted by flowers and other plant parts

nematocide   preparation for eradicating nematodes

nymph   immature feeding stage of some insects

omnivorous   feeding on all kinds of food

ootheca   egg case

oviparous   egg-laying

ovisac   egg sac

parasite   organism dependent on another living organism for nutrition

parasitoid   uses the host for transport and dispersal – not for food

parthenogenic   the ability of some females to reproduce without males

pesticide   preparation for eradicating pests

photosynthesis   process used by green plants to absorb and convert radiant energy to chemical energy

phytotoxic   damaging to plants

predator   organism that captures living prey for food

raptorial   seizing and grasping

saprophyte   organism that feeds on dead plant or animal material

sclerote   hard, asexual resting body produced by some fungi

secretion   substance produced for a particular purpose – protective cover or nectar

sporangium   a structure in which spores are produced

symbiosis   an association of two organisms for mutual benefit

systemic pesticide   a pesticide that enters the sap stream of plants to kill sucking pests

tissue culture   method of propagating plants on nutrient gel under controlled conditions

vector   carrier of a disease

viviparous   gives birth to live young

INDEX

2,2-DPA   260

Acacia fimbriata   45
acacia flea beetle   93
acacia flower gall   38
acacia leaf-spotting bug   59
acacia rust galls   222
Acarina   7, 25
Aceria cynodoniensis   67
Aconophora compressa   152
Acrididae   95
Acrocercops chionosema   100
acrophylla stick insect   124
Acrophylla titan   124
Aethina concolor   91
Agamidae   206
Ageratina adenophora   36
Agriolimacidae   185
Agrobacterium tumefaciens   228
air pollution   248
Alectoria superba   96
Aleeta curvicosta   157
Aleyrodidae   14, 50
algae   28
Alternaria sp.   224
Amblypelta lutescens   58
Amblypelta nitida   58
Amegilla bombiformis   141
Amegilla sp.   141
Amorbus alternatus   62
Amorbus rubiginosus   62
amphibians   215
Amphibolurus sp.   206
amphipods   197
Amylotheca dictyophleba   245
Amyotea hamata   133
Anchiale austrotessulata   97
Ancita marginicollis   161
Animal Kingdom   3–4
Anisoptera   212
Anobiida   99
Anoeconeossa sp.   47
Anomalopus verreauxii   203
Anoplognathus spp.   93
ant lion   188
Anthelidae   110
ants   19, 20–1, 24, 25–6, 111, 115, 176
ant-tended caterpillar   111
Apanteles sp.   119
Aphelenchoides sp.   98
Aphelenchoididae   98
Aphididae   14–15, 48

aphids   14–15, 23, 25, 48, 118, 178
Aphis nerii   48
Apidae   141, 143
Apiomorpha pileata   40, 156
Apolinus lividigaster   127
Arachnura sp.   132
aralia bug   62
Araneidae   131–2
arboreal cockroach   148
Archimantis latistyla   122
Argiope aetherea   131
army worm   83
Arthropoda   6–21
Asclepias sp.   48
Asilidae   213
assassin bug   133
Aterpus griseatus   179
Athoracophoridae   202
Atriplex sp.   48
Austracantha minax   132
Austrotachardia sp.   153
azalea lace bug   64
azalea leaf-miner or roller   83

backhousia root weevil   180
bacteria   25, 27–8, 228, 239
bacterial spot   228
Bactrocera tryoni   102
Balanophora fungosa   246
Balanophoraceae   246
banksia canker   224
banksia hawk moth   106
banksia leaf-curl psyllid   44
banksia mite gall   41
bark bore   120
bark bugs   169
bark cork   242
bark mantis   123
bark miners   24
bark spider   171
bearded dragon   206
bees   20, 24, 141–3
beetles   17, 23–6, 87–94, 99, 127–30, 158–61, 179–81
     bombardier   26, 189
     Christmas   93, 181
     longicorn   25, 160–1
Belostomatidae   214
bigbud   231
Biprorulus bibax   56
bird-dropping spider   132
birds   216
bites   25–6

biting insects   23
bizarre looper   109
blackberry sawfly   86
bladder cicada   144
Blattodea   9–10, 190
blight   223, 224
blister mite   67
blister sawfly   86
blossom caterpillar   74
blossom chaffer beetle   88
blossom spider   131
blue-banded bee   141
blue-spot mantis   123
bombardier beetles   26, 189
borers   24, 80, 159–60, 162–4, 166, 167
boron deficiency   253
boron toxicity   251
Brachychiton discolor   37
brachychiton leaf gall   37
Brachymeria sp.   120
bracket fungus   225
Braconidae   119
brevipalpid mite   65
Brevipalpus sp.   65
Brithys crini   82
broad mite   66
bronze orange bug   57
brown leaf-eating ladybird   90
brown lousy beetle   94
brown moth   110
brown predatory bug   134
brunsfelsia   228
Bucculatricidae   168
bugs   16, 23, 25, 26, 56–64, 133, 134, 169, 214
    lace   63–4
    mealy   15–16, 23, 25, 55, 136, 177
    predatory   26, 133–4
    spittle   13, 23, 135
bulb caterpillar   82
bulb mealy bug   55, 177
bunch mite   65
bunch psyllid   45
burrowing wasp   209
butterflies   18–19, 23, 24, 107–8, 110–11, 120

Caedicia sp.   96
calcium deficiency   254
calico virus   229
callistemon dieback borer   160
Callistemon Gawler hybrid   220
callistemon leaf miner   101
callistemon leaf-rolling thrip   70
callistemon sawfly   84
callistemon speciosus   220
callistemon tip borer   80
callistemon tip bug   60
callistemon white fly   50
Caloptilia azaleella   83

capsule case moth   114
Carabidae   189
Cardiaspina sp.   46
Carsidaridae   44
Caryodidae   149
case moths   81, 113–14
Cassytha sp.   244
casuarina gall   156
caterpillar virus   240
caterpillars   23, 24–5, 72–83, 99, 109, 111, 112, 113, 114, 119, 182, 211
Celaenia excavata   132
centipedes   7, 26, 199
Cephrenes augiades   74
Cerambycidae   160, 161
Cercopidae   135
Ceroplastes rubens   52
Chaetophyes sp.   136
chain scale   51
Chalcididae   120
Charaxes sempronius   108
Chelepteryx chalepteryx   110
chemicals, mixing   32
chewing insects   23–4
Chlaenius flaviguttatus   189
Christmas beetles   93, 217
Chrysomelidae   87–93
Chrysomphalus aonidum   52
Chrysopidae   137
cicada nymph fungus   238
cicadas   13, 23, 25, 144, 157, 192, 238
Cicadellidae   13, 49, 152, 169
Cicadidae   144, 157, 192
cincturing   259
circular black scale   52
cissus leaf miner   100
Clastopteridae   136
cluster caterpillars   76
coastal brown ant   176
Coccidae   15, 51–3, 153
coccids   25, 177
Coccinella transversalis   128
Coccinellidae   17, 127–30
cockroaches   9, 23, 148, 190
Coequosa triangularis   106
Colgar peracutum   49
collar rot   227
Collembola   9, 193
common garden grasshopper   95
common hoverfly   117
common mantis   122
common mealy bug   55
compost beetle   189
compost cockroach   190
control chart   261–2
Convolvulaceae   244
Cordiceps sp.   238
Coreidae   58, 60, 62

Corymbia ptychocarpa   67
Cossidae   166–7
cottony cushion scale   153
couch mite   67
Crambidae   79, 83
crane fly   213
crested grasshopper   96
crickets   12, 23, 24, 26, 146, 191
Crofton weed   36
crown gall   228
crown rot   225
crustaceae   6–7, 197
Cryptes baccatus   154
Cryptocheilus sp.   209
Cryptolaemus montrouzieri   129
Cryptolaemus sp.   51, 153
Cryptophasa sp.   163
Ctenarytaina sp.   46
Ctenomorphodes tessulatus   12
cup moths   111–13
Curculionidae   40, 92, 99, 158–9, 179–80
curl grubs   93, 181
Cuscuta dodder   244
Cuscuta sp.   244
cylindrocladium leaf spot   220
Cylindrocladium sp.   220
Cylindrococcus spiniferus   156
Cyrtophora moluccensis   117
Cystosoma saundersii   144

dahlia ring spot   229
damselflies   9, 212
deficiencies   253–5
Deinopidae   195
Deinopis sp.   195
Delias sp.   110
Dermaptera   11, 170
Dermestidae   130
Deroceras sp.   185
Diaspididae   15, 52–4, 154
Dictyla sp.   63
Didymuria violescens   12
Dindymus versicolor   61
Diplodactylidae   173
Diplodactylus sp.   173
Diplopoda   199, 200
Diptera   17–18, 102–4
Dirioxa pornia   103
dodders   244
domatia   145
Doratifera sp.   112–13
Doratifera vulnerans   113
double-headed hawk moth   106
dragonflies   9, 212
Drosophila sp.   103
Drosophilidae   103

earthworms   6, 201

earwigs   11, 23, 170
echidna   217
Ectobiidae   148
Edusella glabra   91
egg-laying damage   25
ehretia lace bug   63
Elaeocarpus eumundi   68
Elapidae   203
elkhorn fern spore caterpillar   73
Ellipsidion sp.   148
emperor moth   109
Enchesphora lithochlora   77
Endoxyla cinereus   166–7
entomogenous fungus   238
erinose mite   68
Eriococcidae   15, 40, 154, 156
Eriococcus coriaceus   154
Eriophyidae   41, 67–8
Erythraeidae   196
Erythraeus sp.   196
Eucalymnatus tessellatus   53
eucalyptus apiomorphid gall   156
eucalyptus bark borer   167
eucalyptus bubble gall   39
Eucalyptus citriodora   223
eucalyptus leaf spot   221
Eucalyptus miniata   159
eucalyptus nematode or fly gall   39
eucalyptus seedling borer   159
Eucalyptus shirleyi   159
eucalyptus stem and leaf gall   40
eucalyptus tip bug   62
eucalyptus tip gall   40
Eucerocoris suspectus   60
Eucyclodes pieroides   109
Eudocima spp.   72
Eupelmidae   99
Eupelmus sp.   99
Eurycnema goliath   126
Eurytoma sp.   99
Eurytomidae   99
Exocarpos   246
extatosoma leaf insect   125
Extatosoma tiaratum   125

false spider mite   65
false wire worm   189
fasciation   248
Fergusobia sp.   39
Fergusonina sp.   39
Fergusoninidae   39
fern moth borer   164
fern scale   53
field cricket   191
fire ant   176
firetailed resin bee   143
Flatidae   49
flea beetles   92, 93

flies   17–18, 24–5, 26, 36, 39, 50, 102–4, 116–17, 213, 237
flower thrips   69
fluoride toxicity   251
Formicidae   20–1, 115, 176
freezing   249
froghoppers   13, 136
frogmouth   216
frogs   215
fruitflies   25, 102–3
fruit-spotting bug   58
fruit-sucking moth   72
Fulgoroidea   13, 152
fungi   3, 25, 28–9, 220–6, 235–8, 246
fungicides   29
funnelweb spider   194
Furina   203

galls   25, 36–41, 136, 156, 222, 228, 238, 239, 242
garden slug   185
garden snail   185
Gastrodontidae   185
geckos   173
Gekkonidae   173
Geometridae   75, 109
giant mealy bug   136
giant panda snail   149
giant thrip   198
giant water bug   214
giant wood moth   166–7
Glycaspis sp.   39
goliath stick insect   126
Gotra sp.   120
Gracillariidae   83, 100
grasshoppers   11–12, 23, 26, 95–6, 147
green lacewing   137
grevillea bud drop psyllid   42
grevillea flower psyllid   42
Grevillea hodgei   42
Grevillea juncifolia   42
grevillea looper   75
Grevillea rosmarinifolia   42
Grevillea sericea   45
Gryllacrididae   146
Gryllidae   191
Gryllotalpa sp.   191
Gryllotalpidae   191

Hadrogryllacris sp.   146
Halictidae   143
Halticorcus platycerii   92
Haritalodes derogata   79
harlequin bug   61
Harmonia conformis   127
harvest mites   8, 196
hawk moth   76
Hedleyella falconeri   149
Helicarion sp.   149

Helicarionidae   149
Helicidae   185
Heliozela sp.   101
Heliozelidae   101
Helix aspersa   185
Hemibela sp.   114
Hemiberlesia lataniae   154
Herpetogramma licarsisalis   83
Hesperiidae   74
Hexathelidae   194
hibiscus caterpillar   75
hibiscus flea beetle   92
hibiscus flower beetle   91
hibiscus flower fly   102
hibiscus leaf roller   79
hibiscus mealy bug   55
hibiscus woolly psyllid   44
high pH   255
Hippoboscidae   104
honeydew   12–13, 14, 47, 50, 52, 153, 154
hooded brown katydid   147
horticultural problems   257–60
hoverflies   117
Hyalarcta huebneri   81, 114
Hyalinaspis sp.   47
Hylidae   215
Hymenoptera   19–21, 24, 25, 37, 38, 99

Icerya purchasi   153
Ichneumonidae   120
Idolothrips spectrum   198
imperial blue   111
iris bacterial leaf spot   228
iron deficiency   255
irridescent leaf-eating beetle   91
island fruitfly   103
Isopoda   197
Ixodes holocyclus   8
Ixodidae   8

Jacky dragon   206
Jalmenus evagoras   111
jassid   49
jewel spiders   132
jezebel   110
jumping spider   131

Kerriidae   15, 153
key hole wasp   208

lac scale   153
lace bugs   63–4
lace lerp   46
lacerating   23
lacewings   17, 137–40, 188
ladybirds   89–90, 127–30
large flat green scale   51

larger horned citrus bug   56
Lasioderma serricorne   99
Lasioglossum lanarium   143
latania scale   154
Latrodectus hasseltii   195
Lauraceae   244
laurel dodder   244
lawn caterpillar wasp   120
lawn caterpillar   83
leaf rust   222
leaf and twig webber   78
leaf galls   37, 38, 40
leaf miners   24, 83, 98, 100–1
leaf nematode   98
leaf rollers   24, 78
leaf spot   220, 221, 228
leaf thrips   69
leaf tiers   24
leaf-cutter bee   142
leaf-eating ladybird   89
leaf-eating weevil   92
leafhoppers   13, 49, 135–6
legless lizard   203
Lepidoptera   18–19, 24, 25, 72–83, 99, 162–4, 166–8, 182, 211
Lepidota sp.   94
leptospermum gall   40
leptospermum scale   54
lerps   13–14, 23, 46–7
lichen   30, 241
Limacodidae   111–13
liquidambar   242
Lissopimpla sp.   120
Litoria caerulea   215
Litoria sp.   215
loopers   75, 109
long-horned grasshopper   96
longicorn beetles   25, 160–1
long-legged centipede   199
long-tailed spider   132
Lophostemon confertus   38
lophostemon flower and leaf gall   37
lophostemon prickly leaf gall   38
Lophostemon suaveolens   37
Loranthaceae   245
Lycaenidae   111
Lycosidae   194

macadamia leaf miner   100
Maconellicoccus hirsutus   55
magnesium deficiency   254
Mallada signatus   137
mammals   217
manganese toxicity   252
Mantidae   122, 123
mantids   10–11, 26, 122–3
mantis egg beetle   130

mantis egg parasitic wasp   118
mealy bugs   15–16, 23, 25, 55, 136, 177
Megachile sp.   142–3
Megachilidae   142–3
Melaleuca and baeckea psyllid   46
Melaleuca bracteata   223
melaleuca hairy gall   136
melaleuca leaf-spotting bug   60
Melaleuca quinquenervia   136
Melanostoma sp.   117
Meliphagidae   216
Meloidogyne sp.   184
Membracidae   152
Mesohomotoma hibisci   44
Metura elongatus   113
midges   17–18, 26
Millettia pinnata   19
millipedes   7, 199, 200
mining insects   24
Miridae   59–60, 134
mistletoes   245–6
mite-eating ladybird   130
mites   7–8, 23, 25, 41, 65–8, 130, 145, 155, 196
mole cricket   191
Monolepta australis   88
monolepta beetle   88
Monophlebidae   136, 153
Monophlebulus sp.   136
mosquitoes   17–18, 26
moth borers   162–4
moths   18–19, 23–5, 72–83, 106, 109, 111–14, 162–4, 166–8, 182, 211
mottled cup moth   113
mud wasp   208
mud-dauber wasp   207
Musgraveia sulciventris   57
mussel scale   52, 53
Mycorrhiza   238
Myriapoda   7
Myrmeleontidae   188
myrtle rust   223
Myxomycetes   234

Nasutitermes sp.   183
nectar scarab   88
Nematoda   4–5, 25, 98, 184
nematodes   4–5, 25, 39, 98, 184
Neoaratus sp.   213
Neola sp.   110
Neotylenchidae   39
net-casting spider   195
Neuroptera   17, 137–40, 188
nigra scale   153
Nisotra sp.   92
Nitidulidae   91
Noctuidae   72, 82
Notodontidae   110, 211

Notothixos sp.   246
Nuytsia floribunda   243
Nymphalidae   108
Nymphes myrmeleonoides   137
Nymphidae   137

Oecophoridae   114, 163, 164, 167
Oedura sp.   173
Oenochroma vinaria   75
Ogmograptis scribula   168
Oidium sp.   221
Oligochaeta   6, 201
Opodiphthera sp.   109
orange palm dart   74
Orchamoplatus mammaeferus   50
orchard butterfly   107
orchid beetle   87
orchid colour break   230
orchid ring-spot virus   229
orchid snail   185
orchid white fly   50
Ornithomya sp.   104
Orthodera ministralis   123
Osmylidae   138
Oxyopes sp.   131
Oxyopidae   131

palm seed weevil   99
paper wasps   121
Papilio aegeus aegeus   107
Papilionidae   107
Parasaissetia nigra   153
parasites   2
parasitic plants   243–6
parasitised pupa   120
Paropsis sp.   89, 90
pearl scale   154
Pemphigidae   178
Pentatomidae   56, 133, 169
Perga sp.   85
Pergidae   84, 85, 86
pest control   31–3
Pestalotia sp.   224
Pestalotiopsis sp.   223
pesticides   31–3
Petauridae   217
Petaurus breviceps   217
Phaeoseptoria eucalypti   221
Phasmatidae   97, 124–6
phasmids   12, 23, 97, 124–6
Pheidole megacephala   176
Phellineus   225
Phellinius noxius   225
Philagra sp.   135
Phlaeothripidae   70, 71, 198
Phola octodecimguttata   91
phosphorus deficiency   255
phosphorus toxicity   252

Phylacteophaga froggatti   86
Phylidonyris sp.   216
Phyllachora sp.   220
Phyllotocus apicalis   88
Phyllotreta sp.   93
physiological disorders   247–55
Phytophthora parasitica   227
Phytophthora sp.   227
Phytoseiidae   8, 145
Phytoseiulus persimilis   145
phytotoxicity   260
piercing   23
Pieridae   110
pill millipede   200
pimple psyllid   43
pink disease   226
pink wax scale   52, 53
Pinnaspis aspidistriae   53
piper thrips   70
pittosporum thrips   71
Placoasterella sp.   220
plant damage   23–5
plant diseases   27–30
planthoppers   13, 23, 135–6, 152
planting-depth problem   258
plumed scale   53
Podagrion sp.   118
Podargidae   216
Podargus strigoides   216
Podocanthus wilkinsoni   12
Poecilometis sp.   169
Poecilopachys australasia   132
Pogona barbata   206
Polistes sp.   121
Polyphagotarsonemus latus   66
Polyrhachis sp.   115
Pompilidae   209
Pomponatius typicus   60
pot weeds   260
powdery mildew   28, 221
praying mantids   10–11, 122
predatory bugs   26, 133–4
predatory mite   8, 145
Pristhesancus papuensis   133
Procecidochares utilis   36
procession caterpillar   211
Protaetia fusca   88
proteoid roots   242
protista   227, 234
Pseudabispa bicolor   208
Pseudococcidae   15–16, 55, 136, 177
Pseudococcus sp.   55
Pseudoscorpiones   7, 172
pseudoscorpions   7, 172
Psilogramma menephron   76
Psychidae   81, 113–14
Psyllidae   13–14, 39, 42, 44–7
psyllids   13–14, 23, 25, 42–7

Pterygophorus insignis 84
Ptilogyna ramicornis 213
Pulvinaria sp. 51
Pyralidae 77, 78

Quambalaria sp. 223
Queensland fruitfly 102

Rafflesia arnoldi 243
rainforest snail 149
ramularia blight 223
rasping 23
rattler ants 115
Rayieria tumidiceps 59
red spider mite 65, 130
red triangle slug 202
redback spider 195
red-headed flea beetle 92
Red-naped snake 203
red-shouldered beetle 88
Reduviidae 133
reptiles 173, 203, 206
Rhadinosomus lacordairei 159
Rhizobium bacteria 25, 239
Rhytiphora armatula 160
Rhytiphora piperitia 161
rice bubble scale 154
ringbark longicorn beetle 161
robber fly 213
Rodolia cardinalis 153
Ronstar™ damage 260
root aphids 178
root coccids 177
root knot nematode 184
root weevil 179
root-bound plants 258
Ropalidia gregaria 121
rose mosaic virus 229
rot 225–7
Roundup™ 259
roundworms 4–5
rust fungus 25, 222
Rutilia sp. 116

salinity 250
Sarcophaga reposita 117
Sarcophagidae 117
Saturniidae 109
Saunders case moth 113
sawflies 19, 23, 24, 84–6
scale insects 15, 51–4, 115, 153–4
scale wasps 118
Scarabaeidae 88, 93–4, 181
Sceliphron sp. 207
Scincidae 203, 206
Sclerotinia sclerotiorum 226
Sclerotium rolfsii 225
Scolopendridae 199

Scorpiones 7, 172
scorpions 7, 172
scribbly gum moth 168
Scutelleridae 61
Scutigeridae 199
seed wasp 99
seed-eating caterpillar 99
seed-feeders 25
shell lerp 47
shoot-binders 24–5
shoot blight 223–4
shovel nose bug 169
silky case moth 114
Simosyrphus grandicornis 117
sirex wasp 19
skink 206
slaters 6, 23, 197
slime mould 234
slugs 5, 23, 185, 202
snails 5, 23, 149, 185
sod webworm 83
Solanum mauritianum 102, 260
Solenopsis invicta 176
sooty spot 220
Sparassidae 171
Sphaerococcus sp. 136
Sphecidae 207
Sphingidae 76, 106
spider fly 117
spiders 24, 26, 117, 131–2, 171, 194–5
Spilonota constrictana 114
spiny-tail gecko 173
spittle bugs 13, 23, 135
Spodoptera picta 82
spotted lacewing 138
spotted leaf-eating beetle 91
spray burn 260
spray drift 259
sprayers 32–3
springtails 9, 23, 193
St Andrew's cross spider 131
staghorn beetle 181
staghorn fern window-spot beetle 92
stem rot 226
stem rust 222
stem-gumming phytophthora 227
Stenocotis depressa 169
stephania lace bug 63
Stephania spp. 72
Stephanitis pyrioides 64
Stephanitis queenslandensis 63
Sternorryncha 13–16
Stethopachys formosa 87
Stethorus sp. 130
stick insects 12, 23, 97, 124, 126
stick-bound case moth 114
stinging insects 25–6
stinkhorn fungus 235–7

string paper wasp   121
Strophurus intermedius   173
sucking insects   23, 24
sugar glider   217
sugarbag bee   143
sulphur dioxide pollution   248
Syrphidae   117

tabanid fly   103
Tabanidae   103
tachinid fly   116
Tachinidae   116
Tachyglossidae   217
Tachyglossus aculeatus   217
tailed emperor butterfly   108
Talitridae   197
tar spot   220
Tarsonemidae   66
Tecomanthe sp.   66
Tectocoris diophthalmus   61
teddy bear bees   141
Teleogryllus sp.   191
Tenebrionidae   189
tent spider   117
Tenuipalpidae   8, 65
Tephrididae   103
Tephritidae   36, 102
termites   10, 23, 24, 26, 183
Termitidae   183
terrestrial phasmid   126
Tessaratomidae   57
tessellated scale   53
Tetragonula hockingsi   143
Tetranychidae   65
Tetranychus sp.   65
Tettigoniidae   96, 147
Teuchothrips ater   71
Teuchothrips sp.   70
Thaumaglossa nigricans   130
Theridiidae   195
Thomisidae   131
Thomisus spectabilis   131
Thripidae   69
thrips   16–17, 23, 24, 25, 69–71, 198
Thysanoptera   16–17, 25, 69
tick scale   154
ticks   7, 8, 26
Tingidae   63, 64
Tiphiidae   119
Tipulidae   213
tobacco beetle   99
tobacco trees   260
Tortricidae   114
Torymidae   118
toxicity   250–2
transplant problem   258

transverse ladybird   128
tree cricket   146
tree snail   149
tree-dwelling crickets   24
Triboniophorus graeffei   202
Trioza eugeniae   43
Triozidae   43
Tropidoderus sp.   97
tube caterpillar   77
tunnellers   24
twig-binders   24–5
twig-hollowing moth   114

Uromyces sp.   222
Uromycladium sp.   222

Valanga irregularis   95
verrucispora leaf spot   220
Verrucispora proteacearum   220
Vespidae   121, 208
vinegar fly   103
virescence   231
viruses   29–30, 229–30, 240
Viscaceae   246
vitex   91

wallaby fly   104
wasps   19–20, 24, 25, 99, 118–21, 207–10
waterlogging   249
wax wasp   210
webbing borer   167
weevils   25, 92, 99, 158, 179, 180
West Australian Christmas tree   243
whipper snipper 'disease'   259
white caterpillar   182
white clothes beetle   88
white flies   14, 23, 50
white-cheeked honeyeater   216
wild tobacco tree   102
witch's broom mite   155
wolf spider   194

Xanthostemon youngii   223
xerochrysum gall fly   36
Xerohrysum spp.   36, 225
Xyleutes   166–7
Xylotrupes sp.   181

yellow lacewing   137
yellow-tailed black cockatoos   216

zigzag caterpillar   109
zinc deficiency   253
Zonitoides arboreus   185
Zygoptera   212